the nature and application of mathematics

George J. Kertz

The University of Toledo

Goodyear Publishing Company, Inc.
Santa Monica, California

Library of Congress Cataloging in Publication Data

Kertz, George J.
    The nature and application of mathematics.

    Includes index.
    1. Mathematics—1961–   I. Title.
QA39.2.K47   510   78-20979
ISBN 0-87620-614-3

Published by Goodyear Publishing Company, Inc.
Santa Monica, California 90401

Y-6143-5

Current Printing (last digit):
10 9 8 7 6 5 4 3 2 1

Composition and Design: Sandy Bennett

Cover Design: Don Fujimoto
The Arch marking St. Louis as "Gateway to the West."
Chosen for its mathematical features (Chapter 5).

The photo on page 208 is from Bittinger/Crown, Finite
Mathematics: A Modeling Approach, © *1977, Addison-Wesley,*
Reading, Massachusetts, pp. 229. Reprinted with permission.

Printed in the United States of America

To George and Johanna

For all that they contributed

contents

preface

This text was born of the conviction that a mathematics course for liberal arts majors should not, and need not, consist of a number of selected and isolated topics. Instead, by concentrating on the two following objectives, it was felt that the course could be formed into an organized unit with specific goals:

1. To enhance the students' appreciation of the nature of mathematics and of the mathematical method of solving problems in the application of mathematics. After all, these are the basic reasons for the cultural and scientific contribution of mathematics to society.
2. To enhance the students' ability to use mathematics. The liberally educated individual is expected to have some facility in the use of mathematics, especially in view of the increasing variety of applications.

The text began as a rough set of lecture notes for a mathematics appreciation course intended for liberal arts and education majors. The notes evolved through various stages, resulting finally in a multilithed text used campuswide for several years at the University of Toledo. The first of the above goals became the explicit theme—the framework within which the topics for the second goal were presented.

The choice of topics to be included within this framework was influenced by a number of factors: the requirements of the department, suggestions from interested individuals, recommendations of CUPM for "A Course in Basic Mathematics for Colleges" (Mathematics E), and the prejudices of the author. Great care was taken to present the topics in a manner readable to students with a minimum mathematical background. Except for some simple polynomial operations, which can be reviewed if needed, there are no mathematical prerequisites beyond arithmetic for the text. The more sophisticated algebraic operations required in the chapter on analytic geometry are explained when encountered.

There is enough material for the individual instructor to organize the course within the above framework according to his or her own preferences. Chapter 1 and Chapter 4 set the theme for the course and probably should not be omitted (except perhaps for Sections 1.3 and 4.2). However, the material in each chapter can be taught independently of the remaining chapters. The exceptions are Chapter 7 ("Probability II"), which depends on Chapter 2 ("Elementary Probability"), and the equations constructed in Section 4.1, which are graphed in Chapter 5

("Analytic Geometry"). The Instructor's Manual contains several possible course outlines.

The chapter on logic follows the consideration of several mathematical systems, after the students have seen firsthand the need for logic in mathematics. This arrangement seemed preferable to starting the course with logic and ignoring it thereafter. Logic is developed to the point where its use in mathematics can be analyzed in Section 10.9. Otherwise, satisfactory coverage can terminate with Section 10.5 or Section 10.7.

The exercises are grouped in the text where they arise naturally, instead of being collected together at the end of each section. Their location tends to divide the material into convenient one-hour units. An effort was made to keep most of the exercises fairly routine so as not to discourage the student, but rather to reinforce his or her confidence in doing mathematics. The more difficult exercises are marked with an asterisk; the exercises for which the answers are given in the back of the text are marked with a daggar. Two or three sections contain no exercises at all. The material in these sections is intended more as food for thought than for the student's active involvement.

Open-book examinations are recommended at the University of Toledo, because the purpose of the course is to encourage an understanding of mathematical processes, not the memorization of formulas for obtaining solutions. However, the students should be apprised of the danger involved in this type of examination.

In its report to the Mathematical Association of America's Board of Governors, the Committee on New Priorities for Undergraduate Education in the Mathematical Sciences states: "We urge continuing development of courses on the nature of mathematics and its application to society. . . . NSF has recommended more emphasis on science courses for the non-science major. However, we should guard that these are honest courses with real content."[1] Hopefully, this text will serve as a guide for courses in keeping with these two recommendations of the committee.

GEORGE J. KERTZ

---

[1] "Report of the Committee on New Priorities for Undergraduate Education in the Mathematical Sciences," *The American Mathematical Monthly*, November 1974, pp. 984–988.

acknowledgements

To the thousands of students who studied from preliminary versions of this text, particularly those who offered constructive criticism for its improvement;

To their instructors, especially Professors H. Lamar Bentley, Budmon Davis, Charles Davis, Frank Ogg, Richard Shoemaker, and H. Westcott Vayo, the latter for his suggestions and assistance in many aspects of the manuscript development;

To the manuscript reviewers, Professors Gerald Bradley of Claremont Men's College, Burton Rodin of The University of California at San Diego, Robert Stein of California State College at San Bernardino, and James Vick of The University of Texas at Austin, for their valuable suggestions and criticisms (including those which made me cringe!);

To the staff of Goodyear Publishing Company, particularly John Pritchard, Acquiring Editor, for his faith in this project almost from its inception and his guidance throughout its development, and to Linda Schreiber and Laurie Greenstein, Production Editors, for turning an ordinary-looking manuscript into a handsome text;

To Kay Locke for her excellent typing of the several manuscript drafts and for her cheerfulness which made their preparation seem less burdensome;

To my family for the time spent in developing this text that should have been spent with them, and my wife Susan for her support and encouragement throughout, as well as her assistance in proofing the galleys; and

To all who assisted in any way in the preparation of this text, my sincere appreciation.

G.J.K.

note to the /tudent

No one would deny that mathematics is of fundamental importance to our society, but how many people understand why or how mathematics has been able to contribute so much? Such an understanding is not reserved for those individuals who call themselves mathematicians or scientists. It is accessible to every person willing to make the search. The guiding principle is to focus at times on the general picture without undue emphasis on details, except to the extent that such emphasis contributes to the general picture.

By the *general picture* is meant the manner in which mathematical theory is developed and the manner in which the theory is used to solve real-life problems. There one finds a surprising simplicity underlying the powerful structure called mathematics—a simplicity that makes the understanding of the nature of mathematics readily accessible to each interested person. Yet, when enmeshed in the details of this or that mathematical problem, such as manipulating x's and y's in a first course in algebra or proving theorems in high school geometry, one ignores the general scheme of mathematics and often is led to the conclusion that there is nothing simple in mathematics—the forest is lost for the trees!

The purpose of this text is twofold. The first is to bring the forest into focus—to develop an understanding of the nature of mathematics. The second is to consider the development and uses of several areas of mathematics that are applicable in a great variety of situations and that are accessible to you at this stage of your mathematical development. It is my hope that, through the accomplishment of these objectives, this text will enable you to view mathematics as less mystifying, more satisfying, and (this is my fondest hope!) even enjoyable.

The exercises in the text are of two types. Most of them are fairly routine; the examples given and the text discussion should be sufficient to enable you to do most of these. The more difficult exercises are marked with an asterisk (*). It is not expected that anyone will be able to do all of these. But it is hoped that you will enjoy the challenge of trying some and experience the satisfaction that follows when you are successful. The exercises for which the answers are given in the back of the text are marked with a dagger (†).

The final content of the text was greatly influenced by students such as yourself. Any suggestions or comments that you might wish to make would be most welcome.

George J. Kertz
Department of Mathematics
The University of Toledo
Toledo, Ohio 43606

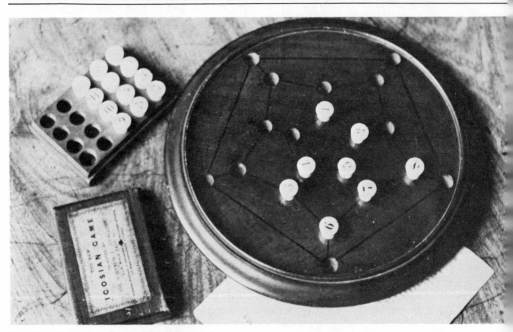

Hamilton's Icosian Game. (Section 1.2.)

graph theory— an introduction to mathematics

1.1 THE BRIDGES OF KÖNIGSBERG

Seemingly trivial circumstances sometimes give birth to great ideas. One such occurrence in the history of mathematics began with seven bridges across the Pregel River in the Prussian city of Königsberg (now the city Kaliningrad in the Soviet Union). The river dividing the city created an island in the interior of the city (Fig. 1.1). A popular interest of the citi-

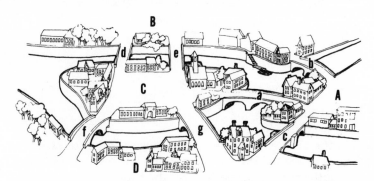

FIGURE 1.1   A, B, C, D denote land areas; a, b, c, . . . , g denote bridges.

zens of Königsberg in the early 1700s was whether a person could start from any of the land areas and take a walk in such a way that he crossed each bridge once and only once. Some thought it could be done, but no one was able to determine how.

The bridges attained a significant position in the history of mathematics when the problem was solved in 1736 by the Swiss mathematician Leonhard Euler (Fig. 1.2). Because one could move around inside the individual land areas in any manner whatsoever, the heart of the problem was in the manner in which the land areas were connected by bridges. Euler recognized this fact and that the relationship between the land areas and the bridges could be represented mathematically by a diagram similar to Figure 1.3, in which the land areas are represented by the points *A*, *B*, *C*, and *D* and the bridges by the lines connecting the points. The question then became whether it was possible to move continuously along the figure in such a way as to traverse each line once and only once.

Leonhard Euler 1707–1783

FIGURE 1.2

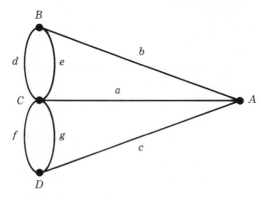

FIGURE 1.3

Euler solved the problem and went on to answer other questions concerning such figures as Figure 1.3. In doing so, he laid the foundation for a new branch of mathematics. Graph theory, the subject of this chapter, is part of this new branch.

Prior to Euler's time, geometry was principally concerned with the magnitude of the various elements of geometric figures. The basic concern of Graph theory is the relationship between the elements of the figures—not the size of any one of them.

Our immediate objective is to see how Euler solved the bridge problem. His solution is applicable to any figure of the type represented by Figure 1.3. We start with some basic terminology.

Points to which or from which lines are drawn are called *nodes*. Lines connecting two consecutive nodes are called *branches*. The *order* of a node is the number of branches that meet at the node. In Figure 1.4, A is a node of order 1, B and D of order 2, and C is of order 3. A node is *odd* if it is of odd order, and it is *even* otherwise. As depicted in Figure 1.4, A and C are odd nodes, B and D are even.

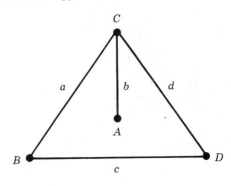

FIGURE 1.4

A *path* is a number of branches traversed consecutively in such a way that no branch is traversed more than once. (Note that *path* is defined in such a way as to describe the type of action considered in the bridge problem—that is, branches are traversed consecutively and no branch is traversed twice.)

To facilitate discussing paths, we shall use the following notation:

BcD   denotes the path that begins at B and traverses branch c to arrive at D.

BcDDdC   denotes the path that begins at B, traverses c to arrive at D, and then traverses d to arrive at C.

BcDDdCCbA denotes a continuation of the previous path from C to A via branch b.

If the number of nodes or branches in a figure could be infinite, the figure could never be described by a single path. Therefore, we restrict the type of figures to be considered by assuming that there are only a finite number of nodes and branches. Such assumptions are called *axioms*. Therefore, we have:

Axiom 1.   The number of nodes and the number of branches are finite.

Also, we want to eliminate from consideration figures that are discon-nected, such as the one represented in Figure 1.5. It would be impossible to describe both parts of this figure by a single path, because there is no way to move from one part of the figure to the other. We can eliminate

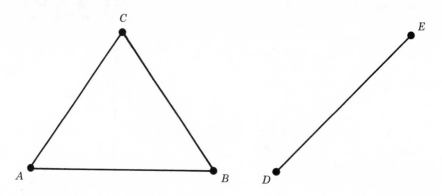

FIGURE 1.5

such figures by requiring that, for any two nodes, there is a path connect-ing them. This gives, as a second assumption,

---

Axiom 2.   For any two nodes A and B, there is a path which begins at A and ends at B.

Because there is no path from A to E, for example, the diagram in Figure 1.5 would be eliminated.

Finally, a *graph* is any collection of nodes and branches that satisfies the two axioms above. The two diagrams in each of Figures 1.6 and 1.7

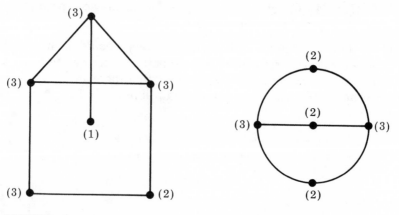

FIGURE 1.6

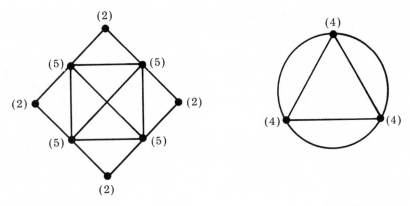

FIGURE 1.7

are graphs. The order of each node is indicated. The sum of the orders of the nodes of the first graph is 14; this sum for the second is 12, for the third 28, and for the fourth 12. Must the sum of the orders of every graph be even? We will return to this question after the following exercises.

EXERCISES

1. Determine the order of each of the nodes in the graphs of Figure 1.8.

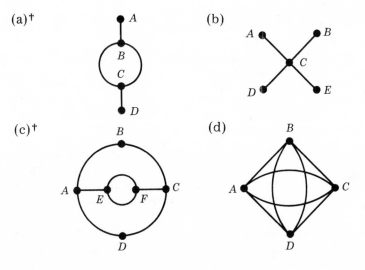

FIGURE 1.8

2.† In each of the graphs of Figure 1.8, which nodes are even and which are odd? What is the sum of the orders of the nodes in each graph?

Note: The more difficult exercises are marked with an asterisk; a dagger indicates that an answer is given in Appendix B.

3.† See which of the graphs in Figure 1.8 you can draw without lifting your pencil from the paper and without retracing any branch.

*4.† The nodes in the two graphs in Figure 1.9 represent six different people. Two nodes are connected by a branch only if the two people they represent are acquainted. In the graph on the left in Figure 1.9, *B*, *C*, and *E* all know each other (they are connected by a triangle).

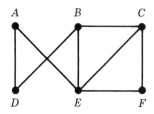
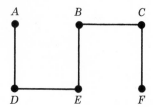

FIGURE 1.9

In the graph on the right in Figure 1.9, none of *A*, *C*, or *E* know each other (there is no branch between any two of them).

Draw at least four more possible "acquaintance" graphs to convince yourself that in any group of six people, there must be three who know each other or three who are complete strangers to each other (no two are acquainted). You can even allow "graphs" in which Axiom 2 does not hold; for example, see Figure 1.10. In this case *B*, *C*, and *F* all know each other; *A*, *E*, and *C* are complete strangers.

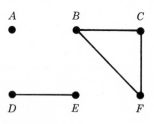

FIGURE 1.10

In the graphs just prior to the exercises, and in the exercises themselves, the sum of the orders of all the nodes was always even. These examples lead us to suspect that this must always be the case—that the sum can never be odd. A little examination reveals that our suspicion is correct.

Every branch contributes two to the sum of the orders—one to the order of the nodes at each of its two endpoints. This means the sum of the orders must be twice the number of branches; therefore, the sum is always even!

We now have another fact, in addition to the axioms, that must be true of every graph. But there is an important distinction. The axioms were basic assumptions required of each graph; that the sum of the orders must be even follows as a consequence of the axioms and definitions. Such consequences are called *theorems*. Like the axioms, theorems are usually stated quite formally:

Theorem 1. The sum of the orders of all of the nodes in a graph is an even number.

What, then, can we say about the number of odd nodes? If there were an odd number of odd nodes, the sum of all the orders could not be even. We have, therefore:

Theorem 2. The number of odd nodes in a graph is even.

Our central question is whether a graph can be completely described by a single path. Because we will want to refer often to such paths, we ought to give them a name. We will call this type of path an *Euler path*.

Graphs for which there are Euler paths—that is, single paths which completely describe the figure—are easily constructed. For the graph in Figure 1.11, *CaAAcBBdCCbA* is one such path.

For the graph in Figure 1.12, *CfDDeCCdDDcBBaAAbC* is one of many such paths.

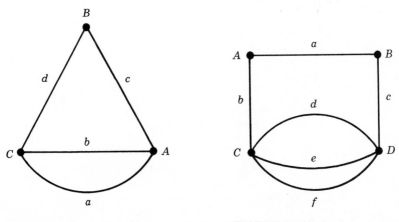

FIGURE 1.11   FIGURE 1.12

Examine the number of times each letter appears in the Euler paths given for these last two graphs. The number of times each letter appears is equal to the order of the corresponding node. For example, in Figure 1.12, node *C* is of order 4 and *C* appears four times in the given path; node *B* is of order 2 and *B* appears two times in the path. This must always be the case because an Euler path must enter or leave each node once for each branch at that node.

Note also that a node which appears *only* in the interior of an Euler path must be even, because such nodes appear only in groups of two. Consequently, if a graph contains odd nodes, these must appear at the beginning and at the end. (Note that this is the case with nodes *C* and *A* in the Euler paths given for the graph in Figure 1.11.) We have then:

Theorem 3.   In a graph with odd nodes, an Euler path must begin and end at an odd node.

In the graph in Figure 1.13, no Euler path is possible, because there are more than two odd nodes. Nodes *B*, *C*, *D*, and *F* are all odd; it would be

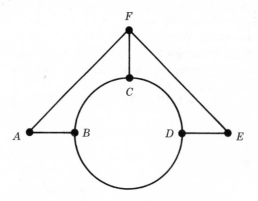

FIGURE 1.13

impossible for an Euler path to either begin or end at each of these four nodes. We can state this fact more generally as:

Theorem 4.   In a graph with more than two odd nodes, no Euler path is possible.

Theorem 4 also solves the bridge problem. In the graph representing the bridges in Figure 1.3, there are four odd nodes. Therefore, no Euler path is possible, which means that it would be impossible to cross each of the bridges exactly once!

With Theorem 4 we have accomplished our objective of seeing how Euler solved the bridge problem. Theorems 5 and 6, which are stated next for the sake of completeness, can also be shown to be valid. Their proofs require a somewhat deeper analysis than those of the first four.

Theorem 5.   In a graph with only even nodes, an Euler path (which must begin and end at the same node) is always possible.

Theorem 6.   In a graph with just two odd nodes, an Euler path that begins at one odd node and ends at the other is always possible.

Theorems 4, 5, and 6 give us enough information to determine whether *any* given graph has an Euler path, and how to construct the path when one exists.

EXAMPLE 1.

Find an Euler path for the graph in Figure 1.14.

All of the nodes are even. Theorem 5 guarantees that there is a path of the required type. One such path is

AaBBfCCeBBbAAcCCdA.

Note that it begins and ends at the same node as the theorem requires.

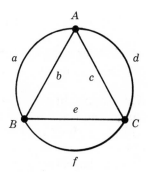

FIGURE 1.14

EXAMPLE 2.

Find an Euler path for the graph in Figure 1.15.

There are two odd nodes. Theorem 6 guarantees that there is a path of the required type. However, the path must begin and end at the odd nodes:

BbDDaAAdBBeCCcD

or

DaAAdBBbDDcCCeB.

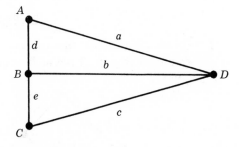

FIGURE 1.15

The path given for Figure 1.14 begins and ends at node *A*. Such paths, that is, paths that begin and end at the same node, are called *circuits*. If a path is both a circuit and an Euler path, it is an *Euler circuit*. One circuit in the graph in Figure 1.16 would be *AbBBcCCdDDaA*. However,

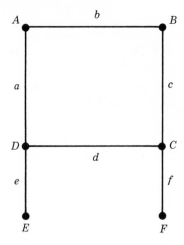

FIGURE 1.16

it is not an Euler circuit, as branches *e* and *f* are not included. In view of Theorems 4, 5, and 6, do you think an Euler circuit can be found for this graph?

While graphs had a rather modest beginning with the bridges of Königsberg, the study of the properties of graphs and of special types of graphs has been developed to the point that applications of graphs are found in a great variety of areas. These include, among others, electronics, chemistry, management science, computer science, sociology, psychology, and transportation. We will consider some of these developments in the next two sections.

EXERCISES

1. On the basis of the theorems of this section, determine whether each graph in Figure 1.17 has an Euler path. Find an Euler path, if there is one.

    (a)†   (b)

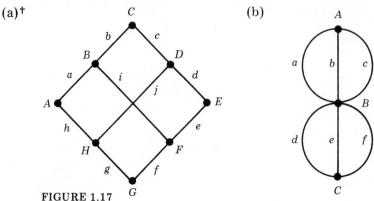

FIGURE 1.17

(c)†

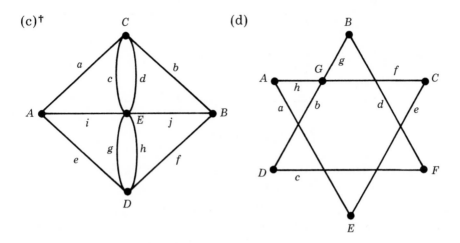

(d)

FIGURE 1.17 (continued)

2. Would Theorems 4, 5, and 6 have helped in Exercise 3 of the last set of exercises?

3.† Which of the paths that you found in Exercise 1 are also Euler circuits? Find a circuit (not necessarily an Euler circuit) that contains all of the nodes for each of the remaining graphs in Exercise 1.

*4. The drawing in Figure 1.18 has only even nodes, but has no Euler path. Why not?

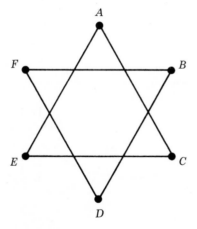

FIGURE 1.18

*5. (a)† A puzzle that was popular some years ago was to draw a continuous line through each of the sixteen line segments in Figure 1.19 without crossing any line segment more than once. (The

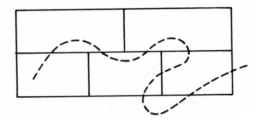

FIGURE 1.19

broken line indicates the type of line which is required.) Show that this is not possible.

(*Hint:* Recall how the figure representing the bridges of Königsberg was constructed on page 2; the land areas were represented by points and the bridges by branches.)

(b)† Construct a puzzle of the above type for which the continuous line can be drawn.

1.2 HAMILTON CIRCUITS

Tom has a paper route represented by the graph in Figure 1.20. The branches represent the streets he must travel, the nodes represent the street intersections. He delivers only on one side of the boundary streets. But on the interior streets, he delivers on both sides and therefore must travel these streets twice. This is reflected by the two branches connecting the

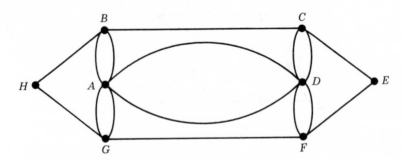

FIGURE 1.20

corresponding nodes; for example, the two branches connecting nodes C and D correspond to the two sides of the street from C to D.

Tom lives at the intersection represented by node A. His best delivery route would correspond to an Euler circuit that begins and ends at A. Theorem 5 guarantees that there is such a route. Can you find one for him?

On Sundays, Tom's father delivers the required number of papers to each intersection by car. Because he does not have to go down any street twice, his route can be represented by the graph in Figure 1.21. He does

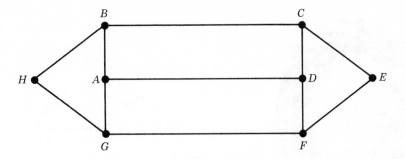

FIGURE 1.21

not have to travel every street, but he does have to get to every intersection. His most desirable route would be a circuit which begins at A, visits each of the other nodes exactly once, and returns to A. He can do this by taking the nodes in the order $ADCEFGHBA$.

The paths which Tom and his father are interested in are both circuits. But note the difference! Tom wants to travel each *branch* exactly once; his father wants to visit each *node* (except for A) exactly once. This latter type of circuit is called a *Hamilton circuit*.

In the graph for the Königsberg bridges (Fig. 1.22), the path $BbAAcDDgCCeB$ is a Hamilton circuit; it arrives at each node exactly once except for the first and last.

(Hereafter, when the branches traveled from one node to another are either obvious or immaterial, we will simplify path notation by just listing

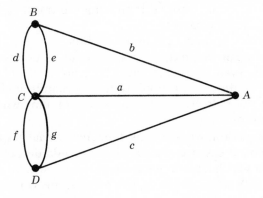

FIGURE 1.22

the nodes in the order in which they are encountered. For example, the circuit just given for the bridge graph would be written *BADCB*.)

The Hamilton terminology began with a puzzle designed by the Irish mathematician William R. Hamilton (Fig. 1.23). The puzzle consisted of a solid wooden dodecahedron as depicted in Figure 1.24. Each of its twenty vertices carried the name of a city of Europe. The object was to find a

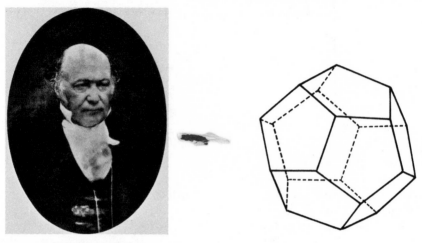

William R. Hamilton (1805–1865)

FIGURE 1.23   FIGURE 1.24

route along the edges which visited each city exactly once. The puzzle does not seem to have been a commercial success; but mathematicians choose the name of its designer to describe the corresponding type of paths in graphs.

Quite likely one would think that it would be just as easy to determine whether a graph has a Hamilton circuit as it is to determine whether it has an Euler circuit. After all, to visit each node exactly once (except for the first) ought not be any more difficult than to travel each branch exactly once. But such is not the case!

We know exactly when a graph has an Euler circuit. Theorems 4, 5, and 6 of Section 1.1 indicate that a graph has an Euler circuit only when all of its nodes are even. Mathematicians have not been able to determine corresponding conditions for a graph to have a Hamilton circuit. The result is that one must resort to trial and error in determining whether a given graph has a Hamilton circuit.

In the next section, we will consider some special graphs in which the question is somewhat easier to answer. If you could determine the exact conditions under which a graph has a Hamilton circuit, you would establish a reputation for yourself among mathematicians who specialize in the study of graphs!

EXERCISES

1. In each of the graphs in Figure 1.25, determine if there is an Euler circuit and if there is a Hamilton circuit.

(a)†   (b)   (c)†

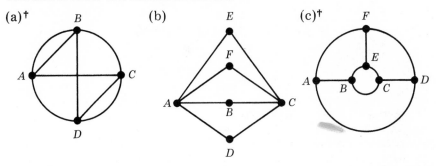

FIGURE 1.25

2. A salesman must call on customers in eight different cities denoted by nodes A through H in the graph in Figure 1.26. The branches correspond to highways connecting the cities.

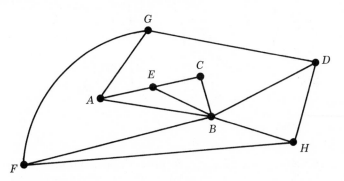

FIGURE 1.26

Can you find a route for him which begins at city A (where he lives), goes to the other cities just once, and returns to city A?

3.† The graph in Figure 1.27 is a two-dimensional version of Hamilton's puzzle. The nodes correspond to the vertices of the dodecahedron and the branches to the edges. Solve the puzzle by finding a Hamilton circuit in the graph.

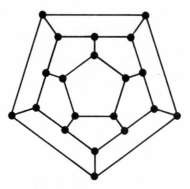

FIGURE 1.27

4. Figure 1.28 is the floor plan for a museum. Construct a graph in which the nodes represent the rooms and any two nodes are connected by a branch if there is a door between the two rooms.

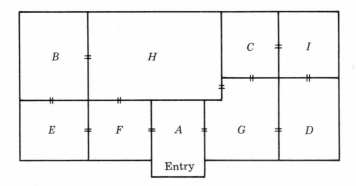

FIGURE 1.28

Any conducted tour of the museum would begin and end at the entry. Would it be preferable for such a tour to be represented in the graph by an Euler circuit or a Hamilton circuit? Can you find such a circuit?

*5. The knight in chess can make two different types of moves; it can move one square horizontally and two squares vertically, or it can move one square vertically and two squares horizontally. All of the possible moves out of (or into) the squares marked S are given in the two drawings in Figures 1.29 and 1.30. There are a total of 336 different moves the knight can make. Is it possible for a knight to start at any position, make all 336 moves without repeating any move, and end up in the position from which it started?

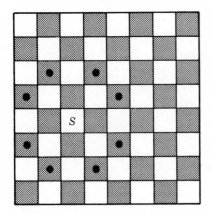

FIGURE 1.29

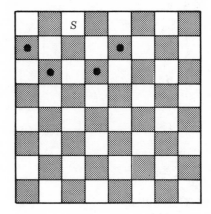

FIGURE 1.30

This question is easily answered using a graph in which the nodes represent the sixty-four squares; two nodes are connected by a branch only if it is possible for the knight to move from one of the corresponding squares to the other. In Figure 1.29, the node for *S* would be of order 8; in Figure 1.30, of order 4. The question then becomes whether an Euler circuit is possible.

(a)† Show that the 336 moves cannot be made in succession.

   (*Hint:* Do not construct the graph, but determine the orders for the nodes representing the squares on the boundary of the board.)

   A "knight's tour" is a sequence of moves around the board which returns the knight to the square at which it began and visits no other square more than once. Such tours that include each of the sixty-four squares correspond to Hamilton circuits. Many of these can be found, if one has the perseverence. Instead, we will consider knights' tours in abbreviated boards.

(b) Find a knight's tour in the 3-square × 3-square board which includes all of the squares except the center (Fig. 1.31).

FIGURE 1.31

(c)† Determine the minimum number of knight's tours which include each square of the 4-square × 5-square board (Fig. 1.32). No square can be included in more than one tour.

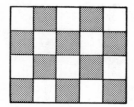

FIGURE 1.32

1.3 MATCHINGS

A company wants to hire an engineer, two computer experts, and a mathematician. They have five applicants, some of whom are qualified for two
of the positions. In Figure 1.33, a branch connects each individual to the
job(s) for which he or she is qualified. An analysis of the graph reveals that
the company is able to fill (or to match) each position with a qualified
person.

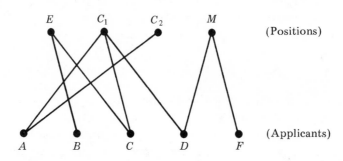

FIGURE 1.33

We will return to the use of graphs for matchings later in this section.
Our present concern is the type of graph used in the matching in Figure 1.33.
The nodes can be divided into two sets, $\{E, C_1, C_2, M\}$ and $\{A, B, C, D, F\}$,
in such a way that nodes in the first set are connected by branches to
nodes in the second set only; and nodes in the second set are connected
only to nodes in the first. The nodes in any path in the graph must alternate between the two sets. Consequently, there can be no Hamilton circuit in the graph if the two sets do not contain the same number of nodes!
This type of two-part graph is called a *bipartite graph*. For a graph
to be bipartite, all of its nodes must be able to be divided into two sets in
such a way that no node is contained in both sets, and the nodes in either
set are connected only to nodes in the other. These sets are called *dividing
sets*. The dividing sets for the graph in Figure 1.33 are the two that were
given, $\{E, C_1, C_2, M\}$ and $\{A, B, C, D, F\}$.

The result of this discussion is stated formally as:

Theorem 7.   In a bipartite graph, no Hamilton circuit is possible if the two dividing sets do not contain the same number of nodes.

Theorem 7 is of no help whatsoever if a graph is not bipartite, or if the dividing sets contain the same number of nodes. It only gives conditions under which we know there is no Hamilton circuit. When these conditions do not hold, we are on our own.

EXAMPLE 1.

Determine if the graph in Figure 1.34 has a Hamilton circuit.

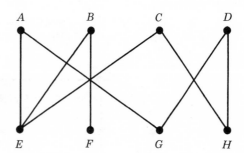

The graph is bipartite, with dividing sets A, B, C, D and E, F, G, H. Because these sets each contain four nodes, Theorem 7 gives us no information.

FIGURE 1.34

Any attempt to find a Hamilton circuit must fail because node F is of order 1. A path can either get into or out of node F; a circuit requires both.

It is easy to see that the graphs in Figures 1.33 and 1.34 are bipartite. But it is not so obvious in the case of the graph in Figure 1.35. However, there is an easy method for checking whether a given graph is bipartite. We illustrate the method for the graph in Figure 1.35.

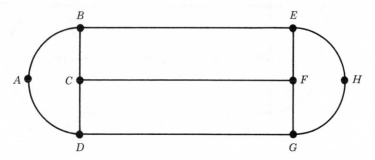

FIGURE 1.35

1. Starting at any node, mark it "1." We start at E (Fig. 1.36).

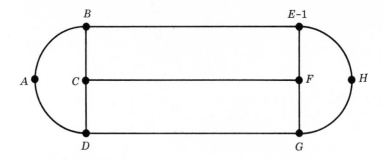

FIGURE 1.36

2. Mark each node directly connected to the one chosen in the first step by "2" (Fig. 1.37).

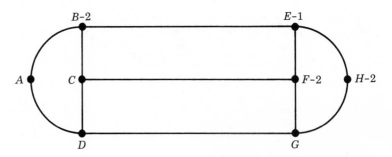

FIGURE 1.37

3. Mark with a "1" each node directly connected by a branch to a node designated "2" in Step 2 (Fig. 1.38).

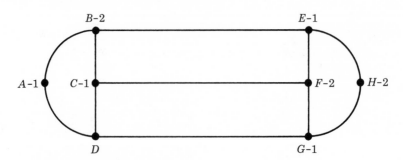

FIGURE 1.38

4. Mark with a "2," each node directly connected by a branch to a node designated "1" in Step 3 (Fig. 1.39).

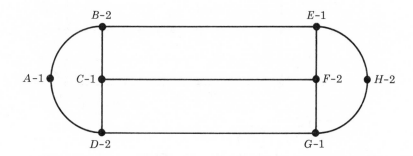

FIGURE 1.39

5. The marking continues until every node has its connected nodes marked with the appropriate number(s). In our illustration, the last node marked was D, and all nodes connected to D are already marked "1."

If each of the nodes is marked with only one number, the graph is bipartite. Furthermore, the 1-nodes and the 2-nodes also make up the dividing sets. In our illustration, the dividing sets are ⅼA, C, E, Gⅼ and ⅼB, D, F, Hⅼ. If one or more nodes are marked both "1" and "2," the graph is not bipartite.

EXAMPLE 2.

Determine whether the graph in Figure 1.40 is bipartite.

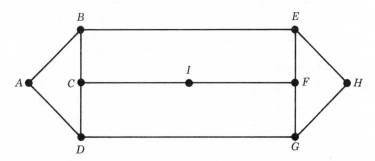

FIGURE 1.40

Starting with node E, we mark the graph as in Figure 1.41. When we consider the nodes connected to B, we mark C with a "1."

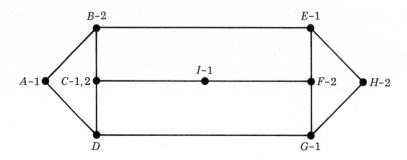

FIGURE 1.41

When we consider the nodes connected to I, we are forced to mark C with a "2" as well. The graph is not bipartite.

Note that Theorem 7 gives us no information about the graphs in Figures 1.39 and 1.41. However, each happens to have a Hamilton circuit.

EXERCISES

1. Determine which of the graphs in Figure 1.42 are bipartite.

(a)†   (b)

(c)†   (d)

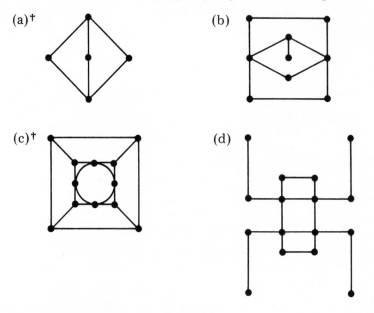

FIGURE 1.42

2. (a)† In which of the graphs in Figure 1.42 does Theorem 7 guarantee that there is no Hamilton circuit?

   (b)† For the remaining graph(s) of Figure 1.42, is there a Hamilton
   circuit?

*3.† If one corner is cut away from a chessboard (Exercise 5, Section 1.2),
   can the knight make a Hamilton circuit? (*Hint:* Consider how Theorem 7 applies.)

*4. The rectangle in Figure 1.43 represents part of a printed electronic
   circuit.

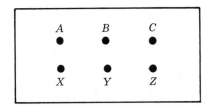

FIGURE 1.43

It is desired to connect terminals A, B, and C to each of the terminals
X, Y, and Z. To avoid short-circuiting, the connections can never
cross each other.

This corresponds to a graph in which each node in $\{A, B, C\}$ is
to be connected with each node in $\{X, Y, Z\}$ by branches which never
cross. At first, the construction of such a graph might seem simple
enough, but it is actually impossible. The proof of this fact is fairly
difficult and will not be attemped here. Instead we will consider one
special case.

In the graph in Figure 1.44, A has been connected to X, Y, and
Z; B has been connected to X and Y.

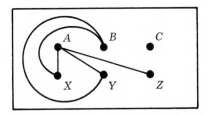

FIGURE 1.44

(a) Connect B to Z by any allowable branch. Then show that it is
impossible to connect C to each of X, Y, and Z in the desired
manner.

(b)   Repeat the procedure in part (a) for two more allowable branches connecting B and Z.

(c)   Can you determine the role that circuits play in your work in parts (a) and (b) that prevents the construction of the required connections with C?

We return now to the bipartite graph at the beginning of this section that joins individuals to the jobs for which they are qualified (Fig. 1.33).

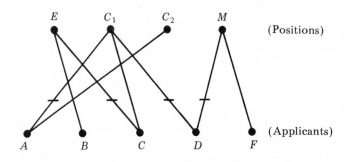

FIGURE 1.45

Removing the branches marked with bars gives us one way in which the available jobs can each be matched with a qualified applicant (Fig. 1.46). Applicant D does not get a job because there were more applicants than jobs.

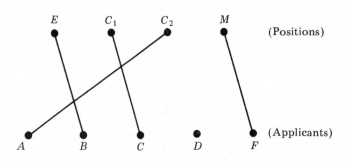

FIGURE 1.46

From the graph point of view, we were interested in matching each node in one dividing set (Positions) with a node in the other dividing set (Applicants) on a one-to-one basis. We obtained one possible matching by eliminating branches from the original graph. The Position nodes are said to be *matched* with Applicant nodes by this procedure.

Note that the Applicant nodes could not be matched with Position nodes for the simple reason that there were not enough Position nodes.

EXAMPLE 3.

The night nursing supervisor of a particular hospital has a group of floating nurses with which to fill vacancies due to absenteeism. One night vacancies occur in the Maternity, Surgery, Cardiac, Nursery, and Urology Units. The graph in Figure 1.47 indicates which of the floating nurses are qualified for these units. Removing the branches with a bar matches each hospital unit with a qualified nurse.

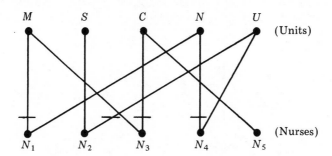

FIGURE 1.47

EXAMPLE 4.

Suppose the graph in Example 3 had joined Units and Nurses in the manner shown in Figure 1.48. The three nodes S, N, and U are collectively joined to only the nodes N_3 and N_4. It is impossible for the three nodes to be matched; consequently, the entire set of Unit nodes cannot be matched.

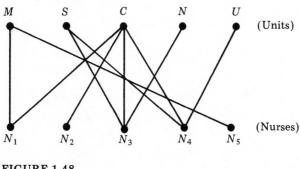

FIGURE 1.48

The procedure used in Example 4 gives a quick way to determine when a match is not possible; every smaller group or subset of the nodes to be matched must collectively be joined to at least the same number of nodes in the other dividing set. If even one subset is not so joined, then no match is possible.

On the other hand, if every subset is so joined, then a match can always be found. This fact is useful in very large matching problems, but will not be used here. In smaller problems, it is easier to examine the graphs than to examine all of the subsets.

EXAMPLE 5.

Construct a matching graph in which every subset of three nodes is collectively joined to at least three nodes, but some subset of four nodes does not have the corresponding property.

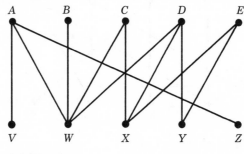

FIGURE 1.49

In Figure 1.49 the possible subsets of three nodes of the top dividing set are $\{A, B, C\}$, $\{A, B, D\}$, $\{A, B, E\}$, $\{A, C, D\}$, $\{A, C, E\}$, $\{A, D, E\}$, $\{B, C, D\}$, $\{B, C, E\}$, $\{B, D, E\}$, and $\{C, D, E\}$. Each of these are joined to the required number of nodes. For example, $\{A, D, E\}$ is joined to $\{V, X, Y\}$. However, $\{B, C, D, E\}$ is joined to $\{W, X, Y\}$—and no other nodes.

EXERCISES

1.† Find a matching for the top dividing set in Figure 1.50.

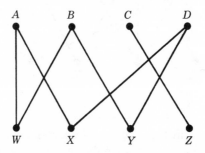

FIGURE 1.50

2. In Figure 1.51, verify that every subset of two nodes in the top dividing set is joined to at least two nodes. Find a subset of three nodes that are not collectively joined to at least three nodes.

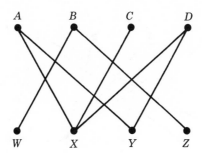

FIGURE 1.51

3.[†] Construct a single matching graph that has all of the following properties (a), (b), and (c):

(a) Each dividing set contains five nodes.

(b) Each subset of three nodes in the top dividing set is joined to at least three nodes of the bottom dividing set. (*Note:* There are ten different subsets of three nodes.)

(c) Some subset of four nodes in the top dividing set is not joined to four nodes of the bottom dividing set.

4. The tradition in a small village is that the boys and girls of the village marry only other citizens of the village. Moreover, relatives are not allowed to marry. Presently there are six boys and six girls considered to be of marriageable age. Hopefully they can be paired in such a way that all can be married.

   The list below gives the girls which each boy is eligible to marry —in that they are not related.

B_1: G_1, G_3
B_2: G_1, G_4
B_3: G_2, G_4
B_4: G_3, G_5, G_6
B_5: G_5
B_6: G_2, G_4

(a) Construct a matching graph in which one dividing set represents the boys, and the other represents the girls. Two nodes are to be joined by a branch if the two corresponding individuals are eligible to marry each other.

(b) Use the graph constructed in part (a) to determine if all twelve individuals can marry without breaking tradition.

5.† The mathematics department of a small college has five faculty members. Their membership in department committees is given by the graph in Figure 1.52. A branch joins two nodes when the corresponding mathematician is a member of the indicated committee.

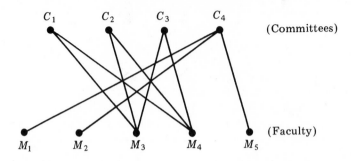

FIGURE 1.52

It is preferred that different committees have different chairmen—that is, no person be chairman of more than one committee. Using the given graph, determine if this arrangement is possible.

*6. One can guarantee that the desired chairmanship arrangement in Exercise 5 is always possible by the following procedure. Pick a number k, where k is not greater than the number of eligible members; require each committee to have at least k members but allow no individual to belong to more than k committees.

Using this procedure, restructure the committees in Exercise 5 so that each committee can have a different chairman. Choose a value of k greater than one.

1.4 THE DUAL NATURE OF MATHEMATICS

Consider again the process by which Euler solved the bridge problem. Every use of mathematics to solve a real-world or physical problem, be it financial, sociological, ecological, engineering, or whatever, involves a similar process. Of course, the details and complexity change.

The two main features in Euler's solution are these:

1. There was obtained a *mathematical description* of the physical situation in the form of the diagram with its nodes, branches, and so forth.
2. In the mathematical structure, or *mathematical system*, of nodes, branches, paths, and so forth (independent of the given diagram), a solution was found—that no figure with more than two odd nodes can be described by a single path.

These two features, a mathematical description and a mathematical system, are basic in every application of mathematics to solving physical problems. These features will be examined in more detail later; as the essential ingredients in the mathematical approach to solving physical problems, they are the central concepts of our study of mathematics.

Mathematics is generally divided into two classifications, *pure* and *applied*, depending upon which of the two central concepts is emphasized. Pure mathematics is primarily concerned with the development of mathematical systems. The term *pure* is adopted by mathematicians to indicate that this type of mathematics is purely theoretical; a more popular term would probably be *theoretical*. Applied mathematics is primarily concerned with obtaining mathematical descriptions of physical situations to answer questions about these situations. It is applied mathematics that each of us encounters daily in our lives.

The reader is cautioned not to infer that pure and applied mathematics are distinct categories with no interaction between the two. As we shall see, by their very nature they are so intimately related that they cannot be separated.

To understand the nature of mathematics, one must realize this dual aspect of mathematics, must be able to distinguish between the two, and must be able to appreciate the relationship between the two. Our purpose is to examine this dual nature of mathematics.

The use of mathematical descriptions will be examined more fully in Chapter 4. But note that the solution of each of the "word" problems in the previous sections began with a mathematical description in the form of a graph. The remainder of this section will consider the mathematical system concept, particularly as we have seen it develop in the previous three sections. Graph theory is a microcosm of the world of pure mathematics in that, as we shall see, the components of any mathematical system have their counterpart in graph theory.

The principal ingredients of every mathematical system are of five types:

1. Undefined terms,
2. Nontechnical terms,
3. Defined terms,
4. Axioms (also called Postulates or Premises), and
5. Theorems.

No system, mathematical or otherwise, can define all of the terms involved in that system without entering into a circle of definitions. You must have encountered this situation in the use of a dictionary. If one wishes to determine the meaning of the word *expound*, he or she is given the word *interpret*; under *interpret* is given *explain*, under *explain* there appears *expound* once again. Every mathematical system also must contain some mathematical terms which are *undefined terms*. In graph theory, *point* and *line* were undefined terms; they were mathematical terms whose meaning was never explained. In fact, at times *lines* were not what we might intuitively have expected them to be.

The *nontechnical terms* are ordinary words without any special meaning in mathematics. Such words might be *an, with, because*, and so forth.

With the undefined terms and the nontechnical terms, the *definitions* of the remaining mathematical terms can be stated exactly. In graph theory, for example, nodes were defined as follows: points to which or from which lines are drawn are called *nodes*. Note the use of the undefined terms *point* and *line*, and the nontechnical terms such as *to, which*, and *called*. Other defined terms include *path, circuit, bipartite graph*, and others. The defined terms were always italicized when their definitions were stated.

The *axioms* (postulates or premises) are the basic assumptions of the system. They are statements about the defined and undefined terms which are accepted as valid without any kind of proof. In graph theory, we encountered just two.

Axiom 1.   The number of nodes and the number of branches are finite.

Axiom 2.   For any two nodes A and B, there is a path which begins at A and ends at B.

Finally, the *theorems* are statements that are logical consequences of the definitions and axioms. The verification that a particular statement is such a consequence—that is, is a theorem—is called a *proof*. The seven theorems of graph theory that we considered are easy to identify. Their proofs were discussed, but were never formally stated as proofs. In later chapters we will encounter theorems whose proofs will be formally stated.

If you took a high school geometry course, you have seen all of these concepts before. At that time, you were involved in pure mathematics, but probably never thought of it as such.

How does the mathematician choose the elements of a particular mathematical system? Strictly speaking, he is free to formulate the system in any manner he chooses as long as it does not contain any contradictions. Usually, however, he is motivated by some particular problem. Euler wanted to solve the bridge problem. The Greeks developed geometry in order to describe the physical world about them. (The word *geometry* comes from two Greek words meaning earth and measure.) The systems that future generations of mathematicians continue to study and develop are the ones that they find interesting or useful for solving physical problems. A lot of mathematics has been developed but long forgotten.

Once the definitions and premises of a mathematical system are established, the development of the system consists primarily in determining the theorems—that is, in determining the logical consequences of the definitions and premises. But it is generally not apparent which statements concerning the elements of a mathematical system will be theorems and which will not—for which a proof can be established and for which there cannot. How then are they arrived at?

They usually begin as educated guesses, arrived at either by some insight into the relationships between the various elements, or just as often by examining some examples of the structure. This is precisely how we arrived at Theorem 1 on page 7 and Theorem 3 on page 8 of Section 1.1 on graph theory! You are asked to guess possible theorems in this manner in the exercises at the end of this section.

The determination of the theorems involves a great amount of effort (and frustration) that is not apparent to the individual who is presented with a developed system in which the theorems are given one after the other, along with their proofs. What is not given in such a presentation are those statements which have been determined not to be theorems. Actually, these "nontheorems" represent a significant part of pure mathematics.

Also, there are questions concerning mathematical structures which have been raised but never resolved. One of these that we encountered in graph theory was, "What are the exact conditions under which a graph has a Hamilton circuit?" These *open problems*, as the mathematicians call them, form a significant part of the work of the research mathematician in the development of mathematics.

We have seen two types of reasoning at work in the development of mathematics: *inductive* and *deductive*. Inductive reasoning is the process of reasoning that arrives at a general conclusion based on multiple observations or multiple occurrences. It is this type of reasoning that is used by the chemist in his laboratory when, after observing that certain chemicals behave in a particular manner during repeated experiments, he concludes that these chemicals will always behave in the same manner in the same circumstances.

It is this type of reasoning that is used by one's employer when, after repeated observations of the quality of the employee's work, he concludes that the employee is superior and therefore should be promoted. It is by this type of reasoning that one arrives at possible theorems by examining examples of a mathematical structure. Inductive reasoning observes particular occurrences and arrives at general principles.

Deductive reasoning, on the other hand, is the process of reasoning that arrives at a conclusion based on an accepted set of premises. It is this type of reasoning that the chemist employs when he predicts the behavior of chemicals based on accepted principles. It is this type of reasoning that is employed in establishing proofs of theorems in mathematics.

No guess ever becomes part of a mathematical structure—becomes a theorem—until it has been established to be a logical consequence of the axioms and definitions. For this reason, mathematics is considered to be a deductive science, as distinct from the other sciences whose basic principles are established mostly by inductive reasoning.

Historically, the deductive method by which mathematical systems are developed stems from the work of the Greek philosophers. These men believed that the meaning of all reality could be arrived at by deductive reasoning from a set of basic principles. That this approach to philosophy should carry over to mathematics was altogether natural because the Greeks, at times, considered mathematics to be intimately related to the meaning of reality.

More precisely, it was the work of Euclid that was most responsible for the deductive method of mathematics today. Not only did his work firmly establish the method in mathematics (although it was probably employed to some extent by prior mathematicians), but it was the importance of his work which preserved the method for future generations of mathematicians.

Consequently, while many mathematical systems are relatively new creations, the tradition that dictates how they are to be developed is centuries old.

This first chapter has had a twofold purpose. The first was to examine, at least in part, the development of an interesting mathematical structure or system. The second was to set the stage for the remainder of the book. Each of the succeeding chapters will be concerned with one or both of the central concepts just discussed, the development of a mathematical structure and/or its use in solving physical problems—pure and applied mathematics. The aspects just discussed concerning the development and use of mathematics will be explored further in a variety of situations.

EXERCISES

In each of the following exercises you are asked to guess a possible theorem for graph theory. This is to be done by examining examples of graphs to determine properties that they have in common. There is no "correct" answer, because any pattern that you recognize as common to the given examples could be a possible theorem.

1. A *tree* is a graph which contains no circuits. Figure 1.53 contains examples of trees. Count the number of nodes and the number of branches in each of these trees.

    (a)   Based on your count, can you guess a theorem which states the relationship between the number of nodes and the number of branches in a tree?

    (b)   Does your guess hold in graphs that are not trees?

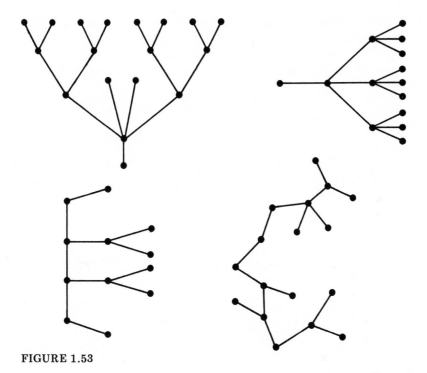

FIGURE 1.53

2. The branches in the graphs of Figure 1.54 never intersect. Such graphs are called *planar*. Count the number of nodes, the number of branches, and the number of regions in these graphs. (The outside is always considered to be one region; the graph in the upper left has three regions.)

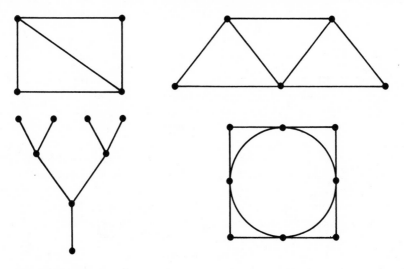

FIGURE 1.54

(a)† Based on your count, can you guess a theorem that states the relationship between the number of nodes, the number of branches, and the number of regions in a planar graph?

(b)† Does your guess hold in graphs that are not planar?

3. A *complete graph* is a graph that contains exactly one branch between every pair of nodes. Count the number of nodes and the number of branches in the complete graphs given in Figure 1.55. Based on your count, can you guess a theorem that states the relationship between the number of nodes and the number of branches in a complete graph?

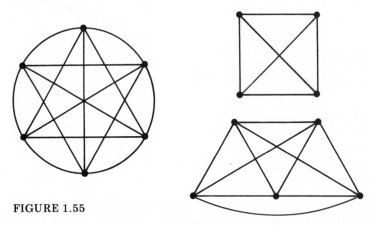

FIGURE 1.55

4.† What type of reasoning (inductive or deductive) did you use in Exercises 1, 2, and 3?

1.5 SUMMARY

The structure of graph theory rests on a number of undefined terms, defined terms, and axioms. Once the foundation is established, the development of the system consists in determining the theorems and in introducing additional defined terms as the study of graph theory takes on new directions.

Our major concern in graph theory was the existence of Euler paths, Euler circuits, Hamilton circuits, and matchings. Euler paths are possible in all graphs which have only even nodes or have no more than two odd nodes; if there are more than two odd nodes, no such path is possible.

Euler circuits are possible in all graphs having only even nodes; they are never possible in graphs having odd nodes. Whether a given graph has a Hamilton circuit is not as easy to determine. Our only theorem on this type of circuit states that if the dividing sets of a bipartite graph do not have the same number of nodes, then no Hamilton circuit is possible.

The question of whether one of the dividing sets of a bipartite graph can be matched with the other dividing set is again not easy to answer, at least for large graphs. However, if one can find a subset of the nodes to be matched that is not matched with a subset of at least the same number of nodes, then no match is possible.

Each application of graph theory to solving a physical problem began with a graph that described the physical situation. The solution, in general, amounted to determining whether there existed an Euler path, Euler circuit, Hamilton circuit, or matching in the describing graph.

The development of the theory and the application of the theory for solving physical problems that we have considered for graph theory illustrate the two major classifications of mathematics—pure and applied. However, one should not conclude that these classifications divide mathematics into two distinct parts with little interaction between the two. We shall see that there is such a close relationship between the two that they can hardly be separated.

Libby Glass Company. Cut lead glass tray. Glass is made from various combinations of materials. However, it is not known why some combinations of materials unite to form glass and others do not. One theory believes that the explanation lies in the probabilistic behavior of the atoms involved.

chapter 2

elementary probability

One of the most glamorous gambling games—with its fast action, bright colors and glittering chrome—is that of roulette. The wheel is spun in one direction, the ball in the opposite direction, and the payoff is made according to the section of the wheel in which the ball falls.

American roulette wheels contain the numbers 0, 00, and 1 through 36 (Fig. 2.1). The numbers 0 and 00 are marked green; the numbers 1 through 36 are one-half red and one-half black. Bets can be made that a red number, a black number, or any one of various combinations of numbers is obtained.

Table 2.1 lists the possible types of bets and the payoff for each type. The manner in which the bet is made is indicated on the roulette layout in Figure 2.2.

FIGURE 2.1

TABLE 2.1   Roulette Bets and Payoffs

Bet title	*Bet description*	*Payoff*
STRAIGHT	Any single number	35 to 1
BLACK or RED	Color	1 to 1
ODD or EVEN	Type of number (Does not include 0 or 00)	1 to 1
LOW or HIGH	(1 to 18) or (19 to 36)	1 to 1
DOZEN	(1 to 12), (13 to 24), or (25 to 36)	2 to 1
COLUMN	(1 to 34), (2 to 35), or (3 to 36) Column	2 to 1
LINE (Six Number)	Six numbers in two adjacent lines	5 to 1
SQUARE	Four numbers forming a square	8 to 1
STREET	Three numbers in a line	11 to 1
SPLIT	Two adjacent numbers	17 to 1
LINE (Five Number)	Numbers 0, 00, 1, 2, or 3	6 to 1

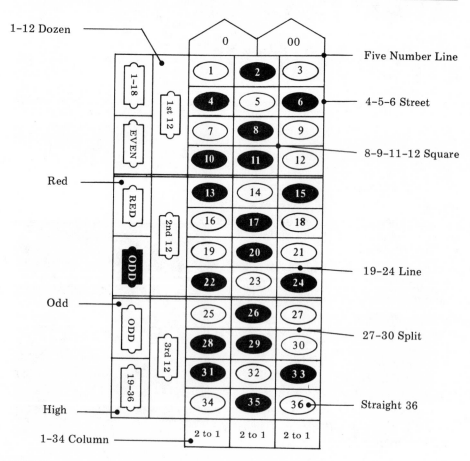

FIGURE 2.2

(It would be a good idea to make a copy of the layout on page 38, as it will be referred to often in this chapter.)

What is the likelihood of winning with any of these bets? Is the pay-off fair in view of the likelihood? Can you expect to make a fortune on roulette if you play long enough?

Questions such as these are easy to answer with elementary probability. By the time we finish this chapter, you will be able to answer them for roulette as well as for other gambling games. But you must suspect already that the green numbers, 0 and 00, give the house some advantage over the gambler.

The first book which treated probability was *The Book on Games* by Girolamo Cardano. (See Fig. 2.3.) This versatile Italian was a physician, mathematician, philosopher, astrologer, and gambler. Having predicted the day on which he would die, he committed suicide on that day so that his prediction would come true!

Cardano's book did not treat probability as a theory in itself, but related probability to problems in gambling. The founders of the theory of probability are generally recognized to be Blaise Pascal and Pierre de Fermat, shown in Figures 2.4 and 2.5,

Girolamo Cardano (1501–1576)

FIGURE 2.3

Blaise Pascal (1623–1662)

FIGURE 2.4

Pierre de Fermat (1601–1665)

FIGURE 2.5

respectively. These two French mathematicians also became interested in probability in connection with problems related to gambling. Today, applications of probability occur in such diverse areas as business administration, industrial engineering, biology, physics, and educational theory.

2.1 PROBABILITY AND SETS

The development of probability theory requires some knowledge of sets. For this reason, we will consider sets prior to discussing probability proper. Sets occur in so many diverse areas of mathematics that they have become part of the common mathematical language along with concepts such as number, angle, circle, and so forth.

A *set* is a collection of objects, all of which have a common property. The collection of words on this page, the collection of books in a library, the collection of teams in the American League, the collection of people in Toledo, the collection of numbers on a roulette wheel—all are examples of sets. The members of a set are called the *elements* of the set; Detroit Tigers is an element of the set of teams in the American League; 7 is an element of the set of roulette numbers.

A bit of notation facilitates working with sets. Uppercase letters, $A, B, C, \ldots$, will be used to denote sets. Lowercase letters, $a, b, c, \ldots$, will be used to denote elements of sets.

Membership in a set is denoted by the Greek letter ϵ (epsilon). To denote, for example, that a is an element of set B, write $a \epsilon B$ (read "a is an element of B"). Similarly, to denote that a is not an element of set C, write $a \notin C$ (read "a is not an element of C"). If A is the collection of all roulette numbers, then $7 \epsilon A$ but $39 \notin A$.

One way to denote the elements of a set is to list all of the elements of the set. This is usually done using braces, as the following examples illustrate. The set of all integers between 0 and 10 is written

$$\{1, 2, 3, 4, 5, 6, 7, 8, 9\}.$$

The set of all vowels is written

$$\{a, e, i, o, u\}.$$

Another way to describe the elements of a set is by use of *set builder* notation. Using this notation, the two sets of the previous paragraph would be written

$$\{x: x \text{ is an integer}, 0 < x < 10\}$$

(read "the set of all x, where x is an integer, x is greater than zero and x is less than ten"), and

$$\{x: x \text{ is a vowel}\}$$

(read "the set of all x, where x is a vowel"), respectively. Again, if

$A = \{x : x$ is an even integer, $0 \leqslant x \leqslant 10\}$

then $A = \{0, 2, 4, 6, 8, 10\}$.

Consider the two sets A, just given, and

$B = \{0, 1, 2, 3, 4, 5, 6, 7, 8, 9, 10\}$.

Every element of A is also an element of B; to describe this relationship between sets A and B, A is said to be a subset of B. Formally, set A is a *subset* of set B if and only if every element of A is also an element of B. To indicate that A is a subset of B, the following notation is used:

$A \subset B$

(read "A is contained in B" or "A is a subset of B").

If B is the set of letters of the alphabet and D is the set of vowels, then $D \subset B$. To denote that a set C is not a subset of a set E, the notation

$C \not\subset E$

(read "C is not contained in E" or "C is not a subset of E") is used.

Two sets A and B are *equal* if and only if they contain the same elements, without regard to the order in which they are listed. If

$A = \{0, 2, 4, 6, 8, 10\}$

and

$B = \{x : x$ is an even integer, $0 \leqslant x \leqslant 10\}$,

then it follows that $A = B$. Note that if $A = B$, then $A \subset B$ and $B \subset A$.

Two special sets are the *empty set* and the *universal set*. The *empty set* is the set that contains no elements. It is denoted by $\emptyset$, a letter of the Danish alphabet, and is considered to be a subset of *every* set; that is, for every set A, $\emptyset \subset A$.

The set that contains all of the elements that enter into a particular discussion is called the *universal set*, denoted by U. If the universal set is not obvious in any particular discussion, then it must be specified.

For every set $A \subset U$, there is associated another set, called the *complement of A* (relative to U), which consists of all those elements of U which are not elements of A. The complement of A is denoted by A' (read "A complement" or "A prime").

If

$U = \{x : x$ is an integer, $0 < x < 20\}$

and

$A = \{1, 2, 3, 4, 5, 6, 7, 8, 19\}$,

then

   $A' = \{9, 10, 11, 12, 13, 14, 15, 16, 17, 18\}.$

Note that $U' = \emptyset$ and $\emptyset' = U$.

   Sets are usually represented pictorially by Venn diagrams. Three examples are shown in Figures 2.6, 2.7, and 2.8. In these diagrams, the universal set is represented by a rectangle; the points interior to the rectangle represent the elements of the universal set. Subsets of the universal set are then represented by circles, or other figures interior to the rectangle; the points inside these figures represent the elements of these subsets.

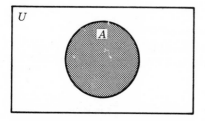

FIGURE 2.6   The points in the shaded area represent the elements of set A.

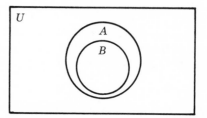

FIGURE 2.7   Representation of B as a subset of A.

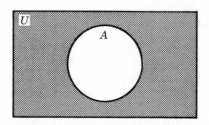

FIGURE 2.8   The points in the shaded area represent the elements of A'.

   The two main operations on pairs of sets are *union* and *intersection*. The *union* of two sets, A and B, is the set that consists of those elements that are either in A or in B, or in both A and B. The union of sets A and B (Fig. 2.9) is denoted by

   $A \cup B$

(read "A union B"); hence

   $A \cup B = \{x : x \in A, \text{ or } x \in B, \text{ or } x \in A \text{ and } x \in B\}.$

Let

$A = \{0, 2, 4, 6, 8, 10\}$
$B = \{1, 2, 3, 4, 5\}$.

Then

$A \cup B = \{0, 1, 2, 3, 4, 5, 6, 8, 10\}$.

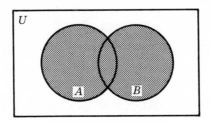

FIGURE 2.9   The points in the shaded portion represent the elements in $A \cup B$.

The *intersection* of two sets A and B is the set consisting of all those elements which are in both A and B (Fig. 2.10). The intersection of A and B is denoted by

$A \cap B$

(read "A intersect B"); hence

$A \cap B = \{x : x \in A \text{ and } x \in B\}$.

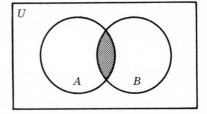

FIGURE 2.10   Representation of $A \cap B$.

If A and B are as in the previous example, then

$A \cap B = \{2, 4\}$.

Note that for every set $A \subset U$,

$A \cap U = A$
$A \cup U = U$
$A \cap \emptyset = \emptyset$
$A \cup \emptyset = A$
$A \cup A' = U$
$A \cap A' = \emptyset$.

EXERCISES

1.† Let $U = \{-6, -5, -4, -3, -2, -1, 0, 1, 2, 3, 4, 5, 6\}$,
$A = \{-6, -4, -2, 0, 2, 4, 6\}$,
$B = \{x: x$ is an integer, $0 < x < 7\}$,
$C = \{x: x$ is an integer, $-7 < x < 7, x$ is a multiple of $3\}$.

Determine:

(a) $A \cup B$
(b) $A \cap B$
(c) $(A \cap B) \cup C$
(d) A'
(e) C'
(f) $A' \cup C'$
(g) $(A \cup C)'$

2. Let $U = \{a, d, i, j, l, m, n, o, p, t, u\}$,
$J = \{j, i, m\}$,
$P = \{p, a, u, l\}$,
$T = \{t, o, m\}$,
$D = \{d, a, n\}$,
$A = \{p, a, t\}$.

Determine:

(a) $P \cap A$
(b) $(P \cap A) \cap D$
(c) $J \cup T$
(d) $(J \cup T)'$
(e) $J' \cap A'$
(f) $P' \cap U$
(g) $\emptyset \cup D$

3.† Let U, A, B, and C be as in Exercise 1 above, and let $D = \{2, 4, 6\}$. Insert a proper symbol from the set $\{\epsilon, \notin, \subset, \not\subset, =\}$ between:

(a) 2 A
(b) D A
(c) A B
(d) $\{3\}$ C
(e) $\emptyset$ C
(f) U B
(g) 3 D

4. Let U, J, P, T, D, and A be as in Exercise 2 above. Insert a proper symbol as in Exercise 3 above.

(a) m $(J \cap T)$
(b) $\{m\}$ $(J \cap T)$
(c) D J'
(d) j A
(e) j A'
(f) $\emptyset$ $(J \cup P)$
(g) P D

5. Indicate which portions of the Venn diagram in Figure 2.11 represent the following sets:

(a)† B'
(b) $(A \cup B)'$
(c)† $(A \cap B)'$
(d) $A' \cup B$
(e)† $A \cap B'$

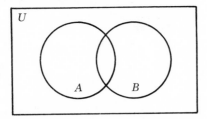

FIGURE 2.11

6. Indicate which portions of the Venn diagram in Figure 2.12 represent the following sets:

(a)† $(A \cup B) \cup C$ (b) $(A \cap B) \cap C$ (c)† $(A \cup B)' \cup C$
(d) $(A \cup B)' \cap C$ (e) $A \cap (B \cup C)$

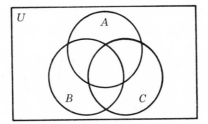

FIGURE 2.12

2.2 SAMPLE SPACES AND EVENTS

The term *experiment* in probability is used in the very broad sense of referring to any observable phenomenon—from the choice of a marriage partner, to the relationship between the planets in the solar system, to the spin of a roulette wheel—as contrasted to its usual reference to a procedure by a scientist in a laboratory. Before determining the probability of any particular outcome of an experiment, one must first be able to determine all of the possible outcomes of the experiment. A set that describes all the possible outcomes of an experiment is the *sample space* of the experiment.

EXAMPLE 1.

The experiment is the roll of a single die. The sample space is therefore the set

$S = \{1, 2, 3, 4, 5, 6\}$

where 1 denotes the outcome of obtaining the side of the die with a single dot, 2 denotes the outcome of obtaining the side of the die with two dots, and so forth.

EXAMPLE 2.

The experiment is the spin of a roulette wheel. The sample space is

$$S = \{0, 00, 1, 2, 3, 4, 5, 6, 7, 8, 9, 10, 11, 12, 13, 14,$$
$$15, 16, 17, 18, 19, 20, 21, 22, 23, 24, 25, 26, 27,$$
$$28, 29, 30, 31, 32, 33, 34, 35, 36\}.$$

We will see later how outcomes such as getting a RED number or a LINE are handled in the sample space.

EXAMPLE 3.

The experiment consists of tossing three coins: a nickel, a dime, and a quarter. The sample space is

$$S = \{(HHH), (HHT), (HTH), (HTT), (THH), (THT), (TTH), (TTT)\}$$

where the first position in each triple indicates the outcome of the nickel, the second position indicates the outcome of the dime, and the third the outcome of the quarter. (HHT) indicates, therefore, that a head is obtained on the nickel and the dime and a tail on the quarter.

EXAMPLE 4.

The experiment consists of a quiz in which five questions are given and a student is to answer any three of the five. The sample space of possible choices of the student is

$$S = \{(123), (124), (125), (134), (135), (145), (234),$$
$$(235), (245), (345)\}$$

where (235), for example, indicates that the student chooses to answer questions 2, 3, and 5.

In connection with Example 1 above, suppose that we are interested in an outcome resulting in an even number. This outcome would correspond to the subset $\{2, 4, 6\}$ of the sample space. Such subsets are called events. Formally, an *event* is a subset of a sample space.

Again, in connection with Example 1, if E is the event that a number greater than 2 is obtained, then

$$E = \{3, 4, 5, 6\}.$$

In Example 2, if E is the event that RED wins, then from the roulette layout at the beginning of this chapter,

$$E = \{1, 3, 5, 7, 9, 12, 14, 16, 18, 19, 21, 23, 25, 27, 30, 32, 34, 36\}.$$

The event E that the LINE consisting of the six numbers 19 through 24 wins would be

$E = \{19, 20, 21, 22, 23, 24\}.$

In Example 3, if E is the event that two heads are obtained, then

$E = \{(HHT), (THH), (HTH)\}.$

EXERCISES

1.† Give the elements of the following events in the sample space of Example 1 of this section:

(a)   An odd number is obtained.
(b)   A number less than 4 is obtained.
(c)   The number 4 is not obtained.

2.   Give the elements of the following events in the sample space of Example 2 of this section:

(a)   ODD wins.
(b)   The 7-8-10-11 SQUARE wins.
(c)   The 1 to 34 COLUMN wins.

3.† Give the elements of the following events in the sample space of Example 3 of this section:

(a)   One tail is obtained.
(b)   At least two heads are obtained.
(c)   No heads are obtained.

4.   Give the elements of the following events in the sample space of Example 4 of this section:

(a)   Question 1 is answered.
(b)   Questions 2 and 3 are answered.
(c)   Question 1 is not answered.

5.† The experiment consists of the roll of a pair of dice, one red and one green.

(a)   Give the sample space where the elements of the sample space are of the form (2,3), which indicates that a 2 was obtained on the red die and a 3 on the green. That is, (2,3) is one of the elements; you are to determine the rest.
(b)   List the elements in the event
(i)   A sum of 5 is obtained.
(ii)   A sum less than 5 is obtained.
(iii)   The sum obtained is an odd number less than 7.

6. Six horses run in a race.

(a)   Give a sample space which gives the possible ways the horses can finish in first and second positions. (Assume that there are no ties, that these are the only two positions of interest, and that at least two horses finish the race.)

(b)   List the elements in the event

(i)   Horse number 1 finished in first position.

(ii)   Horse number 1 finished in first or second position.

(iii)   Horse number 1 did not finish in first or second position.

(iv)   No horse finished in first or second position.

Note that sample spaces and events are the mathematical description of all possible outcomes and of particular outcomes, respectively, for an experiment. Sample spaces and events, in turn, are sets; hence, any further relevant characteristics of the outcome of an experiment must also be expressed in the terminology of sets. In particular, we shall be interested in the manner in which combinations of events are expressed in the terminology of sets.

In Example 1 on page 45, let E_1 be the event "an even number is obtained" and E_2 be the event "a number greater than 3 is obtained." Then

$$E_1 = \{2, 4, 6\}$$

and

$$E_2 = \{4, 5, 6\}.$$

If E_3 is the event "an even number is obtained *or* a number greater than 3 is obtained," then

$$E_3 = \{2, 4, 5, 6\}$$

which is the same as $E_1 \cup E_2$; that is,

$$E_3 = E_1 \cup E_2.$$

In general, the connective *or* between events is expressed as the *union* of the events.

In Example 2 on page 46, let E_4 be the event "the 1-2-4-5 SQUARE wins" and E_5 be the event "the 5-6-8-9 SQUARE wins"; then

$$E_4 = \{1, 2, 4, 5\}$$

and

$$E_5 = \{5, 6, 8, 9\}.$$

The event "the 1-2-4-5 SQUARE wins *or* the 5-6-8-9 SQUARE wins" is

$$E_4 \cup E_5 = \{1, 2, 4, 5, 6, 8, 9\}.$$

Similarly, the connective *and* between events is expressed as the *intersection* of the events. In connection with events E_1 and E_2, if E_6 is the event "an even number is obtained *and* a number greater than 3 is obtained," then

$$E_6 = \{4, 6\}$$

which is the same as $E_1 \cap E_2$.

In connection with events E_4 and E_5, the event "the 1-2-4-5 SQUARE wins *and* the 5-6-8-9 SQUARE wins" is

$$E_4 \cap E_5 = \{5\}.$$

Finally, the event E *not* happening is expressed by E', the *complement* of E with the sample space being the universal set. In Example 4 on page 46, if E is the event "question 1 is answered," then the event "question 1 is *not* answered" is

$$\{(234), (235), (245), (345)\}$$

which is the same as E'.

In Example 3 on page 46, if E is the event "more heads than tails are obtained," the event "more heads than tails are *not* obtained" is

$$E' = \{(HTT), (THT), (TTH), (TTT)\}.$$

EXERCISES

1. Let the sample space be that of Example 4 of this section. Let E_1 be the event "question 3 is answered" and E_2 be the event "question 5 is answered." Write the following events in terms of one of or both of the events E_1 and E_2; then list the elements in each event:

   (a)† Question 3 is answered and Question 5 is answered.
   (b) Question 3 is answered or Question 5 is answered.
   (c) Question 5 is not answered.

2. Let the sample space be that of Exercise 5 of the last set of exercises. Let E_1 be the event "a 1 is obtained on the red die" and E_2 be the event "an even sum is obtained." Write the following events in terms of one of or both of the events E_1 and E_2; then list the elements in each event:

   (a) A 1 is obtained on the red die and an even sum is obtained.
   (b)† A 1 is obtained on the red die or an even sum is obtained.
   (c) An odd sum is obtained.
   (d) A 1 is obtained on the red die and an odd sum is obtained.

3. Let the sample space be that of Exercise 6 of the last set of exercises. Let E_1 be the event "horse number 3 finished in first place," E_2 be

the event "horse number 4 finished in first place," and E_3 be the event "horse number 5 finished in second place." Write the following events in terms of one or more of the events E_1, E_2, and E_3; then list the elements in each event.

(a)   Horse number 3 finished in first place and horse number 5 finished in second place.
(b)   Either horse number 3 or number 4 finished in first place.
(c)†   Horse number 4 did not finish in first place.
(d)   Horse number 3 and horse number 4 finished in first place.
(e)   Horse number 3 finished in first place or horse number 5 did not finish in second place.

2.3   SYSTEM OF PROBABILITY

The system of probability developed here is appropriate only for sample spaces with a finite number of elements, which has been the case for all the sample spaces considered thus far. Other sample spaces may contain an infinite number of elements. For example, the sample space for the experiment of a runner running the mile, where the sample space consists of all the possible times that it might take the runner to finish, would contain an infinite number of elements. In such situations, a system of probability other than that developed here must be used.

Definition.   A basic requirement, along with the sample space, for determining probabilities is a probability function. A *probability function* on a sample space S assigns to each element x in S a number such that the axioms listed below are satisfied. The number assigned the element x is denoted by P(x) and is called the *probability of x*.

Axiom 1.   The probability assigned to each element x in S is a positive number; that is, $P(x) > 0$ for each $x \in S$.

Axiom 2.   The sum of all the probabilities assigned to the elements of S is one.

For example, if the sample space for the roll of a single die (Example 1 of Section 2.2) is $S = \{1, 2, 3, 4, 5, 6\}$ and the assignment of probabilities is

$$P(1) = \tfrac{1}{6} \qquad P(2) = \tfrac{1}{6} \qquad P(3) = \tfrac{1}{6} \qquad P(4) = \tfrac{1}{6} \qquad P(5) = \tfrac{1}{6} \qquad P(6) = \tfrac{1}{6}$$

then the two axioms for a probability function are satisfied.

The probability of the occurrence of any event in the sample space is then defined in terms of the probabilities of the elements in S as follows.

Definition.   For any event $E \subset S$, the *probability of E* is the sum of the probabilities assigned to the elements of E. The probability of an event E is denoted by $P(E)$.

In the roll of a single die, where the event of obtaining an even number of dots less than 5 is $E = \{2, 4\}$, $P(E)$ is given by

$$P(2) + P(4) = \tfrac{1}{6} + \tfrac{1}{6} = \tfrac{1}{3}.$$

What interpretation should be given to the number $\frac{1}{3}$ just obtained for the probability of E? To what extent does it indicate the likelihood of obtaining an even number of dots less than 5 on the next roll of a die?

The probability of an event E indicates the proportion of times that E can be expected to occur if the experiment is performed a great many times. Since the probability of an even number less than 5 being obtained is $\frac{1}{3}$, one would expect the proportion of times that this event occurs in, say 600 rolls of the die, to be about $\frac{200}{600}$ because

$$\frac{200}{600} = \frac{1}{3}.$$

Only to this extent does $\frac{1}{3}$ indicate the likelihood of obtaining a 2 or 4 on any one roll.

The definition above for the probability of an event E does not apply to the empty set, because it does not contain any elements whose probabilities could be added. Because the sample space describes all of the possible outcomes, the empty set corresponds to impossibility. Therefore, it is given probability zero; that is, $P(\emptyset) = 0$.

Notice that the probability of any event E is determined by the manner in which the probability function is determined. For this reason, care must be given to the manner in which the probability function is chosen. If each element in the sample space is as likely to occur as any other, each should be given the same probability. In the case of the die above, presuming the die to be a balanced die, each of the six elements in the sample space was given a probability of $\frac{1}{6}$. In general, if a sample space contains n equilikely elements, each element is given a probability of $1/n$.

Other types of probability functions will be discussed in Chapter 7.

EXAMPLE 1.

Three coins are tossed. What is the probability of getting two or more tails?

The sample space is

$$S = \{(HHH), (HHT), (HTH), (HTT), (THH), (THT), (TTH), (TTT)\}.$$

Since S contains eight equilikely elements, each is given a probability of $\frac{1}{8}$.

The event of getting two or more tails is

$$E = \{(HTT), (THT), (TTH), (TTT)\}.$$

Therefore, the probability of E is $\frac{1}{8} + \frac{1}{8} + \frac{1}{8} + \frac{1}{8} = \frac{1}{2}$

EXAMPLE 2.

What is the probability that RED wins on any spin of the roulette wheel?

The sample space for the spin of a roulette wheel contains the thirty-eight numbers 0, 00, and 1 through 36. Each is equilikely, assuming the wheel to be honest, so each is given a probability of $\frac{1}{38}$. The event RED wins is

$$E = \{1, 3, 5, 7, 9, 12, 14, 16, 18, 19, 21, 23, 25, 27, 30,$$
$$32, 34, 36\}.$$

Therefore, $P(E) = 18 \times \frac{1}{38}$ ($\frac{1}{38}$ added to itself 18 times) or $\frac{9}{19}$.

EXAMPLE 3.

In the Venn diagram in Figure 2.13, the numbers indicate the total probability assigned to the elements of the various subsets. Determine $P(E'_1)$ and $P(E_2 \cup E_3)$.

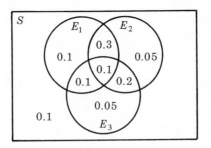

FIGURE 2.13

If we add the probability assigned to the points of E'_1, we get

$$P(E'_1) = 0.05 + 0.2 + 0.05 + 0.1 = 0.4.$$

If we add the probability assigned to the points of $E_2 \cup E_3$, we get

$$P(E_2 \cup E_3) = 0.3 + 0.1 + 0.2 + 0.05 + 0.1 + 0.05 = 0.8.$$

EXERCISES

1.† On the roll of a single die (Exercise 1, page 47), what is the probability that

    (a)   An odd number is obtained?
    (b)   A number less than 4 is obtained?
    (c)   The number 4 is not obtained?

2. Three coins are tossed (Exercise 3, page 47). What is the probability that

    (a)   One tail is obtained?
    (b)   At least two heads are obtained?
    (c)   No heads are obtained?

3. On the spin of a roulette wheel (Exercise 2, page 47), what is the probability that

    (a)† ODD wins?
    (b)† The 7-8-10-11 SQUARE wins?
    (c)† The 1 to 34 COLUMN wins?
    (d)† Why do the events RED wins, BLACK wins, and EVEN wins each have the same probability that ODD wins in part (a)?
    (e)   Why is the probability that any SQUARE wins the same as the probability in part (b) and the probability that any COLUMN wins the same as that in part (c)?

4. On the roll of a pair of dice (Exercise 5, page 47), what is the probability that

    (a)   A sum of 5 is obtained?
    (b)   A sum less than 5 is obtained?
    (c)   The sum obtained is an odd number less than 7?

*5.† Dave and Dennis each roll a die. What is the probability that Dennis rolls a higher number than Dave? (*Hint:* Use the same sample space you used for Exercise 4 above.)

6. A student is given a quiz in which five questions are given, of which he is to answer any three (Exercise 4, page 47). Because the student is well prepared, he makes his choice completely at random. What is the probability that

    (a)   Question 1 is answered?
    (b)   Questions 2 and 3 are answered?
    (c)   Question 1 is not answered?

7.† Suppose that the sample space is $S = \{a, b, c, d\}$ and $P(a) = 0.5$, $P(b) = 0.3$, $P(c) = 0.1$, and $P(d) = 0.1$. Determine the probability of

(a) $E_1 = \{a, c, d\}$ (b) $E_2 = \{a, b, d\}$
(c) $P(E_1 \cap E_2)$ (d) $P(E_1 \cup E_2)$
(e) $\emptyset$ (f) S
(g) Determine $P(E')$ if $E = \{c, d\}$.

8. If the probability assigned to the points of S is as indicated in the Venn diagram in Figure 2.14, determine:

(a) $P(E_1')$ (b) $P(E_1 \cup E_2)$
(c) $P(E_1 \cap E_2)$ (d) $P(E_1 \cap E_2')$

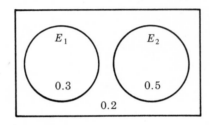

FIGURE 2.14

9.† If the probability assigned to the points of S is as indicated in the Venn diagram in Figure 2.15, determine:

(a) $P(E_1 \cup E_2)$ (b) $P(E_1 \cap E_2)$
(c) $P(E_1' \cap E_2)$ (d) $P[(E_1 \cap E_2)']$

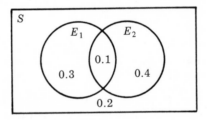

FIGURE 2.15

Now that the definitions and axioms of probability have been established, we can examine their logical consequences—the theorems of probability.

For example, what can we say about the probability of S? Because S is also an event, its probability is defined to be the sum of the probabilities of all of its elements. By Axiom 2, this sum is one. We have, therefore

---

Theorem 1.   If S is the sample space, $P(S) = 1$.

Or consider the events E_1 and E_2 depicted in the Venn diagram in Figure 2.16, in which the sample space is the Universal set. What can we say about the probability of $E_1 \cup E_2$? The probability of the union would

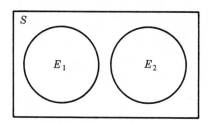

FIGURE 2.16

be the sum of the probabilities assigned to the points in the union. Because E_1 and E_2 have no points in common, we would obtain the same result by adding the probabilities assigned to the points in E_1 to those assigned to the points in E_2.

For example, if the experiment consists of the roll of a single die, if E_1 is the event "a number less than 3 is obtained," and if E_2 is the event "a number 4 or greater is obtained," then

$$S = \{1, 2, 3, 4, 5, 6\}$$
$$E_1 = \{1, 2\}$$
$$E_2 = \{4, 5, 6\}$$
$$E_1 \cup E_2 = \{1, 2, 4, 5, 6\}$$

and

$$E_1 \cap E_2 = \emptyset.$$

If each element in the sample space is given a probability of $\frac{1}{6}$, then

$$P(E_1) = \frac{2}{6} = \frac{1}{3}$$

and

$$P(E_2) = \frac{3}{6} = \frac{1}{2}.$$

Also $P(E_1 \cup E_2) = \frac{5}{6}$, which is the same as $P(E_1) + P(E_2)$.

(If E_1 and E_2 had points in common, that is, $E_1 \cap E_2 \neq \emptyset$, a different result could be expected. The reader is asked to examine the probability of $E_1 \cup E_2$ under these circumstances in the exercises.)

The results of the above discussion can be summarized in Theorem 2.

Theorem 2.   If E_1 and E_2 are events in S such that $E_1 \cap E_2 = \emptyset$, then $P(E_1 \cup E_2) = P(E_1) + P(E_2)$.

Events E_1 and E_2 that have no elements in common, that is, $E_1 \cap E_2 = \emptyset$, are said to be *mutually exclusive*. Mutually exclusive events represent outcomes which cannot happen together. Getting a 2 and getting a 3 on a single roll of a die, or a person getting married and not getting married on a particular day, are examples of mutually exclusive events. Owning a sports car and running out of gas are not mutually exclusive—they can happen together. Theorem 2 states that the probability of either of the mutually exclusive events occurring is the sum of their individual probabilities

For any event E, $E \cap E' = \emptyset$. That is, E and its complement are mutually exclusive. Therefore, we can take E_1 in Theorem 2 to be E and E_2 to be E' to get

$$P(E \cup E') = P(E) + P(E').$$

But $E \cup E' = S$; therefore, this last equation can be written

$$P(S) = P(E) + P(E').$$

Because $P(S) = 1$, we get

$$1 = P(E) + P(E')$$

which proves

Theorem 3.   For any event $E \subset S$, $P(E) + P(E') = 1$.

This last theorem is useful for calculating $P(E')$ when $P(E)$ is either known or is relatively easy to determine, and vice versa. For example, we know from earlier in this section that the probability of getting a 1 or 2 on the roll of a single die is $\frac{1}{3}$. It follows by Theorem 3 that the probability of not getting a 1 or 2 is $1 - \frac{1}{3} = \frac{2}{3}$.

EXAMPLE 4.

Determine in two ways the probability of getting two heads and one tail on the toss of three coins.

As before, the sample space is

$$S = \{(HHH), (HHT), (HTH), (HTT), (THH), (THT), (TTH), (TTT)\}$$

in which each element is given a probability of $\frac{1}{8}$. First, let E be the event of getting any combination other than two heads and one tail. Then,

$$E = \{(HHH), (HTT), (THT), (TTH), (TTT)\}$$

and $P(E) = \frac{5}{8}$. The event we are interested in is E'; by Theorem 3

$$\frac{5}{8} + P(E') = 1$$

or $P(E') = \frac{3}{8}$.

The second method, using the definition of the probability of an event is already familiar. The event of getting two heads and one tail is the subset

$$\{(HHT), (HTH), (THH)\}.$$

Its probability is the sum of the probabilities assigned to the three elements of this subset, which is again $\frac{3}{8}$.

Our last theorem in this section is concerned with the relationship between the probabilities of two events when one is a subset of the other. (See Fig. 2.17.) Since the probability assigned to the points in E_2 includes

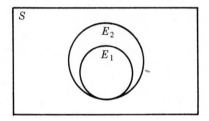

FIGURE 2.17

the probability assigned to the points in E_1, we would expect $P(E_2)$ to be greater than $P(E_1)$. Theorem 4 states that this relationship does hold. In this case a formal proof is given.

Theorem 4.   If E_1 and E_2 are events in S such that $E_1 \subset E_2$, then $P(E_1) \leqslant P(E_2)$.

PROOF.

Let E be the set of those elements of E_2 that are not in E_1 (Fig. 2.18). Then $E_2 = E_1 \cup E$ so that

$$P(E_2) = P(E_1 \cup E). \tag{2.1}$$

But E_1 and E have no elements in common; Theorem 2 tells us that $P(E_1 \cup E) = P(E_1) + P(E)$. Substituting into Equation (2.1) gives

$$P(E_2) = P(E_1) + P(E). \tag{2.2}$$

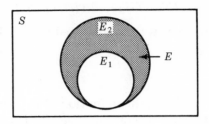

FIGURE 2.18

Axiom 1 and the definition of the probability of an event guarantee that $P(E) \geqslant 0$. Therefore, $P(E_1) + P(E) \geqslant P(E_1)$. From Equation (2.2), it follows that $P(E_2) \geqslant P(E_1)$.

EXAMPLE 5.

Why is the following impossible?

$$E_1 = \{a, d\}, \quad E_2 = \{a, b, c, d\}, \quad P(E_1) = 0.7, \quad P(E_2) = 0.5.$$

Since $E_1 \subset E_2$, Theorem 4 states that $P(E_1)$ must be less than or equal to $P(E_2)$; therefore, it is impossible for $P(E_1)$ to be 0.7 and $P(E_2)$ to be 0.5. The "smaller" set must have the smaller probability.

Theorem 4 is not particularly helpful for calculating probabilities, but does assist in the interpretation of probability values.

By taking the set E_2 in Theorem 4 to be the sample space S, it follows that for each $E \subset S, P(E) \leqslant P(S) = 1$; that is, probabilities never get larger than 1.

Because some element of S must occur when the experiment is performed and $P(S) = 1$, a probability of 1 indicates certitude. As events get "closer" to S, their probabilities increase (by Theorem 4) and the events become more likely to occur—the closer to 1 the probability of an event, the more likely it is that the event will occur.

In the opposite direction, as events get "closer" to $\emptyset$ their probabilities decrease—the closer to 0 the probability of an event, the less likely it is that the event will occur. Also, by taking the set E_1 in Theorem 4 to be the empty set, it follows that for each event E, $0 \leqslant P(E)$. Consequently, $0 \leqslant P(E) \leqslant 1$ for every event E.

Recall the major ingredients given in Chapter 1 for mathematical systems: undefined terms, nontechnical terms, defined terms, axioms, and theorems. In our system of probability, *number* was an undefined term; the *probability of an event* and *mutually exclusive* events were among the defined terms. There were only the two axioms, both stated on page 50. After the statement of the definitions and axioms, the development of the system (apart from any applications) amounted to determining and proving the theorems. Note that each application of probability began with a mathematical description of all possible outcomes in the form of a sample space and of particular outcomes as subsets of the sample space.

EXERCISES

1.[†] Suppose the sample space is $S = \{a, b, c, d\}$ and $P(a) = 0.5$, $P(b) = 0.3$, $P(c) = 0.1$, and $P(d) = 0.1$. If $E = \{a, d\}$, find $P(E')$ by two different methods.

2. You ask a cashier for change for a half a dollar and request no pennies. Assuming that the manner in which she makes the change is purely arbitrary, determine in two ways the probability that the change she gives you will include just one quarter.

3. Why are each of the following situations impossible?
   (a)[†] $S = \{\alpha, \beta, \gamma\}$ and $P(\alpha) = 0.2$, $P(\beta) = 0.3$, $P(\gamma) = 0.4$.
   (b) $E_1 = \{0, 1, 2, 4\}$, $E_2 = \{0, 1, 2, 4, 6\}$, $P(E_1) = 0.8$, and $P(E_2) = 0.5$.
   (c)[†] For some event E, $P(E) = 0.3$ and $P(E') = 0.8$.
   (d) For some event E, $P(E) = 1.5$.
   (e)[†] $E_1 = \{a, b, c\}$, $E_2 = \{d, e\}$, $E_3 = \{a, b, c, d, e\}$, $P(E_1) = 0.3$, $P(E_2) = 0.5$, and $P(E_3) = 0.9$.
   (f) $E_1 = \{\gamma, \beta\}$, $E_2 = \{\beta, \gamma, \delta\}$, $P(E_1) = 0.3$, $P(E_2) = 0.4$, and $P(E_1 \cap E_2) = 0.5$.

4.[†] Consider the relationship between the sample space S and events E_1 and E_2 in Figure 2.19. Can you write $P(E_1 \cup E_2)$ in terms of $P(E_1)$,

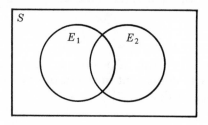

S

E_1 E_2

FIGURE 2.19

$P(E_2)$, and $P(E_1 \cap E_2)$? (Note that Theorem 2 does not apply since E_1 and E_2 do have points in common; the result here is sometimes referred to as the *Addition Rule*.)

5. Given that $P(E_1) = 0.3$, $P(E_2) = 0.5$, and $P(E_1 \cap E_2) = 0.2$, use the result of Exercise 4 as well as the theorems of probability to determine:
   (a) $P(E_1')$
   (b)[†] $P(E_1 \cup E_2)$
   (c) $P[(E_1 \cup E_2)']$
   (d)[†] $P(E_1 \cup E_1')$

6. In the game Over-and-Under, a pair of dice is rolled and one can bet whether the sum of dots showing on the two dice is OVER 7, UNDER 7, or EXACTLY 7. Let E_1 be the event OVER 7 wins, E_2 the event UNDER 7 wins, and E_3 the event EXACTLY 7 wins.

    (a)† Using the sample space for the roll of a pair of dice, determine $P(E_1)$, $P(E_2)$, and $P(E_3)$.
    (b) If you bet that OVER 7 wins, write the event E that the house wins in terms of E_2 and E_3.
    (c)† Use Theorem 2 to determine the probability that the house wins when you bet that OVER 7 wins.

*7. Use the result of Exercise 4 to determine the probability that, in the spin of a roulette wheel, the 1-2-4-5 SQUARE wins or the 5-6-8-9 SQUARE wins.

2.4 PROBABILITIES AND ODDS

Odds give a mathematical description of the likelihood of an event occurring as compared to the likelihood of it not occurring. Probability enters in a natural way.

The *odds in favor of an event E occurring* are simply the quotient

$$\frac{P(E)}{P(E')}$$

[read "$P(E)$ to $P(E')$"] or what is the same thing

$$\frac{P(E)}{1 - P(E)}.$$

Odds are generally given in terms of positive integers. These are obtained by writing the odds as an equivalent quotient of two positive integers with no common factor greater than 1.

EXAMPLE 1.

What are the odds in favor of obtaining "snake eyes" (a one on each die) on the roll of a pair of dice?

Because the probability of obtaining snake eyes is $\frac{1}{36}$, the odds in its favor are

$$\frac{\frac{1}{36}}{\frac{35}{36}} = \frac{1}{35}.$$

The odds are then said to be 1 to 35 in favor of the occurrence of snake eyes, and are written 1:35.

_____

The *odds against an event E occurring* are given by the quotient

$$\frac{P(E')}{P(E)} \quad \text{or} \quad \frac{1 - P(E)}{P(E)}.$$

In Example 1 above, the odds against obtaining snake eyes are

$$\frac{\frac{35}{36}}{\frac{1}{36}} = \frac{35}{1}$$

or 35:1.

The odds against an event will always be the reverse order of the odds in favor of the event.

EXAMPLE 2.

What are the odds against picking the correct answer in a five-answer multiple-choice question if the choice is a pure guess?

The probability of guessing correctly is $\frac{1}{5}$; the probability of guessing wrong is $\frac{4}{5}$. The odds against answering correctly are

$$\frac{\frac{4}{5}}{\frac{1}{5}} = \frac{4}{1}$$

or 4:1.

EXAMPLE 3.

What are the odds in favor of a SQUARE winning in roulette?

Because the probability of a SQUARE winning is $\frac{4}{38} = \frac{2}{19}$, the odds in favor are

$$\frac{\frac{2}{19}}{\frac{17}{19}} = \frac{2}{17}$$

or 2:17.

The odds against a SQUARE winning are then 17:2.

_____

A *fair bet* is defined to be a bet which is in proportion to the odds. Accordingly, from Example 1, a fair bet on obtaining snake eyes should be in the ratio of 1 to 35—$1.00 to $35.00, $.50 to $17.50, $.10 to $3.50, and so forth.

According to the table at the beginning of this chapter, a winning SQUARE pays 8 to 1. A winning $2.00 bet would receive $16.00. According to Example 3, if the bet were fair, the payoff should be $17.00.

EXAMPLE 4.

What should the payoff be for a $1.00 COLUMN bet in roulette in order that the bet be fair?

The probability that a COLUMN wins is $\frac{6}{19}$ (Exercise 3, page 53). The odds in favor are then

$$\frac{\frac{6}{19}}{\frac{13}{19}} \text{ or } 6:13.$$

We want to determine the number x for which the ratio $1:x$ is equal to $6:13$; that is, for what x does

$$\frac{1}{x} = \frac{6}{13}?$$

Multiplying each side of this equation by $13x$, we get

$$13 = 6x,$$

and then, by dividing each side by 6, we get

$$\tfrac{13}{6} = x.$$

The payoff should be $2\frac{1}{6}$ dollars, or about $2.17.

Note that the actual payoff is listed as $2.00 in the table at the beginning of this chapter.

EXAMPLE 5.

Pat bets Dan 25¢ even that in the toss of three coins he will obtain at least two heads. Is this a fair bet?

The probability of obtaining at least two heads is $\frac{1}{2}$ (Exercise 2, page 53). Therefore the odds in favor of such an outcome are

$$\frac{\frac{1}{2}}{\frac{1}{2}} = \frac{1}{1}.$$

The bet is fair—it is in the same ratio as the odds.

The examples and exercises of this section should lead you to suspect that casino gambling games are not fair to the player—at least in the mathematical sense of the word. If one placed hundreds of bets under such circumstances, he could expect to lose in the end.

In Example 4, the odds in favor of COLUMN winning were 6:13. In 1,900 COLUMN bets, the players would be expected to win about 600 times, the house 1,300 times. If all of these bets were $1.00 bets, then the house would collect $1,300 from the losers and would pay $1,200 to the winners. The house would gain a profits of $100.

However, it is rare indeed that a business is established solely for the pleasure and/or profit of its customers. If all bets were fair, the casino could only expect to break even—with no money for its expenses or profit.

EXERCISES

1.† The probability that an event will occur is 0.35.
   (a)   What are the odds in favor of its occurrence?
   (b)   What are the odds against its occurring?

2.   What are the odds in favor of obtaining a number greater than 4 on the roll of a single die?

3.† (a)   What are the odds in favor of EVEN winning in roulette?
   (b)   What would a fair payoff be for a $1.00 bet?

4.   In Over-and-Under (Exercise 6, page 60), winning with OVER 7 and UNDER 7 each pay even money, and EXACTLY 7 pays four-to-one. Determine whether or not these are fair.

5.† The odds in favor of Sweet Evening Breeze winning a race are 4:7. What would a fair payoff be for a $2.00 bet?

6.   Tom bets Jim $2.00 that he will obtain a picture card (ace, king, queen, or jack) on a single draw of a card from an ordinary deck of fifty-two playing cards. How much money must Jim put up to make the bet fair?

*7.† Ed hits the bull's-eye at the archery range about half the time.
   (a)   What are the odds in favor of him hitting the bull's-eye on three consecutive shots?
   (b)   Ed bets Eric 10¢ to Eric's 70¢ that he will hit the bull's-eye on each of his next three shots. Is this a fair bet?

*8.   The odds in favor of an event E occurring are 3:5. Can you determine $P(E)$? (*Hint:* For some x, $3/x \div 5/x = 3/5$; consider the relation between $3/x$ and $5/x$.)

2.5 SUMMARY

The calculation of probabilities (for experiments with a finite number of possible outcomes) begins with a sample space, which is a description of all of the possible outcomes. Probabilities are then assigned to each element in the sample space. Finally, for a given event E, $P(E)$ is determined by adding the probabilities assigned to the elements of E.

The theorems of probability give us some formulas which give additional ways of determining $P(E)$. These are:

$$P(E) + P(E') = 1$$
$$P(E_1 \cup E_2) = P(E_1) + P(E_2) - P(E_1 \cap E_2).$$

If E_1 and E_2 are mutually exclusive, then $E_1 \cap E_2 = \emptyset$ and the last equation above becomes

$$P(E_1 \cup E_2) = P(E_1) + P(E_2), \text{ if } E_1 \cap E_2 = \emptyset.$$

Using these formulas to calculate the probability of an event is sometimes easier than adding up the probabilities assigned to the elements in the event under consideration.

$P(E)$ is always a number between 0 and 1. It guarantees nothing about E happening when the experiment is performed once. Rather it is a long-range indicator; it gives the fraction of times one would expect E to occur if the experiment is performed a large number of times. However, the closer to 1 that $P(E)$ is, the more likely it is that E will occur in a single experiment; the closer to 0, the less likely E will occur.

Odds, while not the same as probabilities, are determined using probabilities. The odds in favor of an event E occurring are given by the quotient $P(E)/P(E')$; those against E occurring are given by $P(E')/P(E)$. A fair bet, then, is a bet which is in the same ratio as the odds.

"Mathematics is the queen of the sciences and the theory of numbers is the queen of mathematics."
                    CARL F. GAUSS (1777–1855)

An IBM System/360 Model 50 installation. Magnetic core storage is located in the dark cabinets on either side of the operator's console. The arithmetic and logic circuits are housed in the cabinet directly behind the operator's console. (See Section 3.7.)

chapter 3

the natural numbers

Arithmetic, algebra, and geometry are the types of mathematics which most of you have previously encountered. How do these fit into the general scheme of mathematical systems?

As indicated in Section 1.4 on the structure of mathematical systems, Euclidean geometry itself forms a mathematical system. The usual course in high school geometry is devoted to the study of the development of this particular system.

To understand the role of arithmetic and algebra, it is necessary to consider the *real number* system. The real numbers can be divided into several classes:

1. The *natural numbers* (also called the positive integers), which are the counting numbers: 1, 2, 3, 4, 5, ...
2. The *integers*, which are the natural numbers, the negatives of the natural numbers, and zero: . . . , $-4, -3, -2, -1, 0, 1, 2, 3, 4, ...$
3. The *rational numbers*, which are all numbers which can be written as the quotient of two integers. For example,

$$3/2, -7/4, 12/-5, 0 = 0/3, 1 = 1/1, \text{ and } -6 = -6/1$$

are all rational numbers.

4. The *real numbers*, which are the rational numbers and numbers such as $\sqrt{2}$, $\sqrt{3}$, and π which cannot be written as the quotient of two integers. (Numbers such as $\sqrt{2}$, $\sqrt{3}$, and π are called *irrational numbers*.)

Note that the integers contain the natural numbers, the rational numbers contain the integers, and the real numbers contain the rational numbers.

Arithmetic is concerned with calculations and manipulations of numbers within the system of real numbers. Algebra is essentially the same, except that in algebra some of the quantities are unknown or variable quantities, represented by x, y, or other symbols. The "modern math" movement is, to a great extent, an attempt to develop an understanding of arithmetic and algebra from this point of view, rather than learning rote manipulations that provide no understanding of the rationale behind the manipulations.

In the application of mathematics, the real numbers outdistance every other mathematical system in both frequency and variety of applications. Some applications are mundane, some are quite esoteric. The development of the sciences and of engineering is due, to a great extent, to the availability of the real numbers for stating and solving the problems in those areas.

The usual method of developing the system of real numbers is to start with the development of the system of natural numbers, then that of the integers, followed by the system of rational numbers. On this foundation, the system of real numbers is developed.

The full development of the system of real numbers is much too lengthy and complicated to be presented in full detail at this point. We shall, however, consider the axioms, definitions, and some theorems of the system of natural numbers to see how the development of the real-number system begins.

3.1 AXIOMS FOR THE NATURAL NUMBERS

The mathematical system that has stood the longest test of time is undoubtedly the natural number system. The permanence of the natural numbers is due to the fact that they are the counting numbers; and man, from the earliest times, found it convenient to count objects. Just like the farmer today, the primitive shepherd kept record of his flock by counting the animals. The symbols for, and names of, the various numbers have changed, but the counting concept remains the same.

It is interesting to note that, in keeping with the name *natural*, these are the first number concepts learned by little children when they begin to talk.

In contrast to their usage, the formalization of the natural numbers as a mathematical system is relatively new. The axioms for the system, which we will give below, were first presented in 1889 by an Italian mathematician, Guiseppe Peano (Fig. 3.1). He obviously did not create the natural numbers at that time. His purpose was to establish a rigorous mathematical

Guiseppe Peano (1858–1932)

FIGURE 3.1

foundation for the set of natural numbers (that is, as a deductive system based on a set of axioms) that already were well known. This foundation consists of a few basic properties of the natural numbers—properties that are almost taken for granted by anyone familiar with the natural numbers:

Axiom 1.   There is a first natural number, namely 1.

Axiom 2.   For each natural number, there is a next larger one, or a "successor."

Axiom 3.   Given any two different natural numbers, their successors are different.

These three axioms require the existence of all the known natural numbers. They require that 1 have a successor; this number is 2. Then 2 is required to have a successor; this number is 3. Then 3 must have a successor, 4; etc., etc., etc.

Axiom 4.   If 1 is in a set S and if, whenever n is in S, the successor of n is in S, then S contains every natural number.

This axiom may seem a bit mysterious at first, but with a little thought its meaning becomes fairly obvious. Suppose a set S satisfies the two conditions of the axiom:

1. 1 is an element of S;
2. Whenever n is in S, then the successor of n is also in S.

We are given that 1 is in S, but is 2 in S? Because 1 is in S, Condition 2 (with $n = 1$) tells us that 2 is in S.

Is 3 in S? Because 2 is in S, Condition 2 (with $n = 2$) tells us that 3 is in S. Then Condition 2 (with $n = 3$) tells us that 4 is in S. Continuing in this manner, we can determine that 5 is in S, 6 is in S, 7 is in S, and so forth.

In fact, we could, with considerable patience and perseverence, show that any given natural number, no matter how large, is also in S. Consequently, if a set S satisfies the two conditions above, S contains all of the natural numbers.

The genius of Peano was in the realization that, along with the definitions of addition and multiplication, all of the properties of the natural numbers can be deduced from these few basic properties. As was pointed out earlier, Peano did not create the natural numbers at the time that he established the axioms. What he did accomplish was the establishment of the natural numbers as a deductive mathematical system.

He defined the sum and the product of two natural numbers in terms of successors. His definitions are fairly technical and will not be considered here, but they give the procedure for finding sums and products from the most elementary

$$1 + 1 = 2$$

and

$$2 \times 2 = 4$$

to those involving the largest natural numbers imaginable.

As was pointed out in the beginning of this chapter, the development of the real-number system begins with this foundation for the natural numbers. On these few axioms and definitions rests one of the most awesome structures developed by man, for it is hard to imagine any area of concerted human endeavor that is not dependent upon the real numbers in one way or another!

3.2 SOME PROPERTIES OF THE NATURAL NUMBERS

In this section, we shall consider some of the well-known properties of the natural numbers that follow from the axioms and definitions. These properties are the statement of the theorems of the system. The proofs of these theorems will not be given. They depend upon fairly sophisticated use of the fourth axiom. The important thing is to appreciate that they follow as consequences of the axioms and definitions.

What are the properties of the natural numbers? We illustrate some of these properties using the natural numbers 3, 11, and 24.

1. $3 + (11 + 24) = (3 + 11) + 24$
    (The grouping of three numbers for the purpose of addition is immaterial: the *Associative Law of Addition*.)
2. $3 + 11 = 11 + 3$
    (The order in which two numbers are added is immaterial: the *Commutative Law of Addition*.)
3. $3 \times (11 \times 24) = (3 \times 11) \times 24$
    (The grouping of three numbers for the purpose of multiplication is immaterial: the *Associative Law of Multiplication*.)
4. $3 \times 11 = 11 \times 3$
    (The order in which two numbers are multiplied is immaterial: the *Commutative Law of Multiplication*.)
5. $3 \times (11 + 24) = (3 \times 11) + (3 \times 24)$
    (The order in which multiplication and addition are performed among three numbers as indicated is immaterial: the *Distributive Law*.)

The theorems which follow state that these properties hold for all natural numbers.

Theorem 1.   (The Associative Law of Addition) If ℓ, m and n are any natural numbers, then $\ell + (m + n)$ $= (\ell + m) + n$.

When adding three natural numbers, we can add the first number to the sum of the second two: $\ell + (m + n)$; or, we can determine the sum of the first two, and then add that result to the third number: $(\ell + m) + n$. The Associative Law of Addition states that in either case, we always get the same sum.

EXAMPLE 1.

By Theorem 1, we know that

1. $(2 + 4) + 7 = 2 + (4 + 7)$
2. $3 + (a + 5) = (3 + a) + 5$
3. $(a + b) + 6 = a + (b + 6)$

whenever a and b are natural numbers.

Theorem 2.   (The Commutative Law of Addition) If m and n
are any natural numbers, then $m + n = n + m$.

The Commutative Law of Addition states that we always get the
same sum no matter in which order two numbers are added. You know
from your own experience that this is always the case: that $3 + 2$ is the
same as $2 + 3$. The point here is that this feature of addition is a consequence of the axioms and definitions for the natural numbers. Not all
operations have this commuting feature. If you need convincing, try passing through a doorway and shutting the door; then reverse the procedure!

EXAMPLE 2.

By Theorem 2, we know that

1. $3 + 7 = 7 + 3$
2. $a + 4 = 4 + a$
3. $(a + b) + 9 = 9 + (a + b)$

whenever a and b are natural numbers.

EXAMPLE 3.

Using the theorems of this section, verify that

$$a + (b + c) = (a + c) + b$$

whenever a, b, and c are natural numbers.

We will start with the left side of the given equation and, using the
two theorems just given, rewrite it to obtain the right side.

By the Commutative Law of Addition, $(b + c) = (c + b)$ so that

$$a + (b + c) = a + (c + b).$$

Then, by the Associative Law of Addition, the right side of this last
equation can be written

$$a + (c + b) = (a + c) + b.$$

Consequently,

$$a + (b + c) = (a + c) + b.$$

EXERCISES

1. Suppose a, b, and c are natural numbers. By which theorems of this section (give the theorem by name rather than by number) do we know that

    (a)† $2 + a = a + 2$?
    (b)† $3 + (a + 9) = (3 + a) + 9$?
    (c) $c + 1 = 1 + c$?
    (d) $5 + (a + b) = (a + b) + 5$?
    (e)† $(a + c) + 2 = a + (c + 2)$?
    (f) $(b + 4) + 1 = b + (4 + 1)$?
    (g)† $12 + 6 = 6 + 12$?
    (h) $(a + b) + c = a + (b + c)$?

2. Suppose a, b, and c are natural numbers. Using only the theorems of this section, verify that:

    (a)† $a + (b + c) = (b + a) + c$.
    (b) $a + (2 + b) = (a + b) + 2$.
    (c)† $(3 + a) + 5 = a + (3 + 5)$.
    (d) $2 + (b + c) = c + (2 + b)$.

Theorems 3, 4, and 5 state the remaining basic properties of the natural numbers that were referred to in the beginning of this section. Theorems 3 and 4 state properties for Multiplication which are similar to those stated by Theorems 1 and 2 for Addition. Theorem 5 combines the two operations.

Theorem 3.   (The Associative Law of Multiplication) If ℓ, m and n are any natural numbers, then $\ell \times (m \times n) = (\ell \times m) \times n$.

Theorem 4.   (The Commutative Law of Multiplication) If m and n are any natural numbers, then $m \times n = n \times m$.

Theorem 5.   (The Distributive Law) If ℓ, m and n are any natural numbers, then $\ell \times (m + n) = (\ell \times m) + (\ell \times n)$.

EXAMPLE 4.

If a, b, and c are natural numbers, we know that

1. $4 \times (c \times 7) = (4 \times c) \times 7$ by Theorem 3.
2. $3 \times (c + 2) = (c + 2) \times 3$ by Theorem 4.
3. $2 \times (a + b) = (2 \times a) + (2 \times b)$ by Theorem 5.

EXAMPLE 5.

Suppose that a and b are natural numbers. Using the theorems of this section, verify that

$$3 \times (a + b) = (a \times 3) + (b \times 3).$$

Starting with the left side again, but in a more concise format:

$3 \times (a + b) = (3 \times a) + (3 \times b)$ by the Distributive Law.

$(3 \times a) + (3 \times b) = (a \times 3) + (b \times 3)$ by the Commutative Law of Multiplication.

Therefore, $3 \times (a + b) = (a \times 3) + (b \times 3)$.

EXAMPLE 6.

Under the same conditions as Example 5, verify that

$$5 \times (a + b) = (5 \times b) + (5 \times a).$$

Using the format of Example 5,

$5 \times (a + b) = (5 \times a) + (5 \times b)$ by the Distributive Law.

$(5 \times a) + (5 \times b) = (5 \times b) + (5 \times a)$ by the Commutative Law of Addition.

Consequently, $5 \times (a + b) = (5 \times b) + (5 \times a)$.

We have considered only the beginning of the development of the natural-number system. However, our purpose is not the full development of the system, but only an appreciation of how the system develops. Do not allow the details to obscure the overall picture. Note that the axioms express the essential features of the natural numbers; the other properties of the natural numbers follow as theorems.

The remaining sections of the chapter consider some unsolved problems concerning the natural numbers and some applications in mathematical games and computer science.

EXERCISES

1. Suppose a, b, and c are natural numbers. By which theorems of this section (give the theorem by name rather than by number) do we know that

   (a)† $a \times 2 = 2 \times a$?
   (b)† $3 \times (b \times 4) = (3 \times b) \times 4$?
   (c)† $a \times (c + 6) = (a \times c) + (a \times 6)$?
   (d) $(2 \times b) \times c = 2 \times (b \times c)$?

(e) $(a \times c) + (a \times b) = a \times (c + b)$?
(f)† $(5 + b) \times c = c \times (5 + b)$?
(g) $5 \times (9 + b) = (5 \times 9) + (5 \times b)$?
(h) $(17 \times 3) \times a = 17 \times (3 \times a)$?

2. Suppose a, b, and c are natural numbers. Using the theorems of this section, verify that:

(a)† $4 \times (a \times b) = (a \times 4) \times b$.
(b) $6 \times (b + c) = (b \times 6) + (c \times 6)$.
(c)† $(c \times 12) \times a = (c \times a) \times 12$.
(d) $(a \times c) + (b \times 3) = (3 \times b) + (c \times a)$.
(e)† $a \times [b \times (c + 2)] = [a \times (b \times c)] + [a \times (b \times 2)]$.

3. Suppose u, v, and w are natural numbers. Show that each of the following is equal to $u \times (v + w)$.

(a) $(v + w) \times u$ (b)† $(v \times u) + (u \times w)$
(c) $(u \times w) + (u \times v)$ (d) $(w \times u) + (v \times u)$

*4.† Can the following statement be a theorem in the natural-number system: If ℓ, m, and n are any natural numbers, then

$$\ell + (m \times n) = (\ell + m) \times (\ell + n).$$

3.3 PRIME NUMBERS

The previous sections were concerned with the foundations of the natural numbers. Further studies of the natural numbers have taken several directions, some of which have resulted in unsolved or *open* problems. As indicated in Chapter 1, these are statements concerning the natural numbers that neither have been verified to be valid nor have been proven to be invalid in the system.

Our purpose in this section and Section 3.4 is to examine some directions that the study of the natural numbers has taken and to point out some unsolved problems that have resulted. You will find that the problems are simply stated and easy to understand. However, to prove or disprove them is not so simple. Mathematicians have worked on their solution for years, some of the problems for hundreds of years. If they are solved, the solutions may well come from someone other than a professional mathematician, because a completely new approach to the problems may be required.

The first direction of investigation we will consider is that of the prime numbers. First, some basic definitions are required.

If m, n, and k are natural numbers such that $m \times k = n$, then m is said to be a *divisor* (or *factor*) of n. If m is a divisor of n, then n is said to be a *multiple* of m. For example, since

$$2 \times 3 = 6$$
$$5 \times 21 = 105$$
$$13 \times 12 = 156,$$

2 is a divisor of 6 and 6 is a multiple of 2; 5 is a divisor of 105 and 105 is a multiple of 5; 13 is a divisor of 156 and 156 is a multiple of 13.

Every natural number n has at least two divisors, itself and 1. Also, any natural number can have many divisors. For example, 1, 2, 3, and 6 are all divisors of 6; 1, 3, 5, 7, 15, 21, 35, and 105 are all divisors of 105.

A natural number is *even* if it is a multiple of 2; natural numbers that are not even are called *odd*. The numbers 2, 12, 158, and 9,320 are all even numbers; 1, 9, 17, and 255 are all odd.

A natural number that is greater than 1 and has no divisors other than 1 and itself is called a *prime number*, or simply a *prime*. There is only one even prime number. Can you determine what it is? Some examples of prime numbers are 2, 3, 5, 7, 11, 13, 19, 23, and 29. The divisors of a natural number n, other than itself, come from the numbers less than or equal to $n/2$. This means the larger natural numbers have many more candidates for divisors than do the smaller. The number 29, for example, has as potential divisors 1, 2, 3, 4, . . . , 13, 14; 1,477 has 1, 2, 3, 4, . . . , 722, 723. Because of this greater abundance of possible divisors, one might expect that there are no primes among the very large natural numbers. But such is not the case. Euclid constructed a proof that verified that there is no largest prime number; consequently, there must be infinitely many primes.

In general, there is no easy way to determine if any given natural number is a prime number. Two easy rules for eliminating the possibility for any given n to be a prime (except $n = 2$ and $n = 5$) are:

1. If the last digit of n is 0, 2, 4, 6, or 8, then n is not a prime. Why?
2. If the last digit of n is 5, then n is not a prime. Why?

A general method for determining if a given natural number n is a prime number (unless n is so large as to make the method impractical) is illustrated below. The method, called the Sieve of Eratosthenes after the Greek mathematician Eratosthenes (266–194 B.C.), is based on two facts:

1. If m is not a divisor of n, then no multiple of m is a divisor of n.

For example, 2 is not a divisor of 105, so neither can 4, 6, 8, or any other multiple of 2 be a divisor of 105. Likewise, 7 is not a divisor of 206, so neither can 14, 21, 28, or any other multiple of 7 be a divisor of 206.

2. If n is not a prime number, n must have a divisor (other than 1) which is less than or equal to the square root of n.

For example,

3 is a divisor of 9, and $3 \leqslant \sqrt{9} = 3$;
2 is a divisor of 36, and $2 \leqslant \sqrt{36} = 6$;
5 is a divisor of 225, and $5 \leqslant \sqrt{225} = 15$;
9 is a divisor of 108, and $9 \leqslant \sqrt{108} \approx 10.8$.

The verification that these two facts are theorems in the natural number system is given in the exercises.

We shall illustrate the sieve method by verifying that 233 is a prime number. Because $15^2 = 225$ and $16^2 = 256$,

$$15^2 < 233 < 16^2$$

or

$$15 < \sqrt{233} < 16.$$

Therefore, if 233 has a divisor other than 1 and itself, one such divisor must be less than or equal to 15. We list the natural numbers 2 through 15.

②	6	10	14
3	7	11	15
4	8	12	
5	9	13	

The first number in this listing, 2, is not a divisor of 233. Consequently, no multiple of 2 can be a divisor of 233. Therefore, we eliminate 2 and every multiple of 2.

~~②~~	~~6~~	~~10~~	~~14~~
3	7	11	15
~~4~~	~~8~~	~~12~~	
5	9	13	

The next number appearing is 3. By division, we can verify that 3 is not a divisor of 233. So we can eliminate 3 and every multiple of 3.

~~②~~	~~6~~	~~10~~	~~14~~
~~③~~	7	11	~~15~~
~~4~~	~~8~~	~~12~~	
⑤	~~9~~	13	

The next number remaining is 5. By division, we can verify that 5 is not a divisor of 233. We therefore eliminate 5 and any multiple of 5 that may still remain in the listing.

This process is repeated with 7, 11, and 13 in the order that they appear. In this manner, all possible divisors of 233 that are greater than 1 and less than or equal to 15 are eliminated.

$$
\begin{array}{cccc}
\cancel{2} & \cancel{6} & \cancel{10} & \cancel{14} \\
\cancel{3} & \cancel{7} & \cancel{11} & \cancel{15} \\
\cancel{4} & \cancel{8} & \cancel{12} & \\
\cancel{5} & \cancel{9} & \cancel{13} &
\end{array}
$$

We conclude that 233 is a prime number. Note that division is performed only with prime numbers in this process.

As a second example, we shall determine whether 1,001 is a prime number. Because $31^2 = 961$ and $32^2 = 1,024$,

$$31 < \sqrt{1,001} < 32.$$

We list all of the natural numbers from 2 through 31

$$
\begin{array}{cccccc}
\cancel{2} & \cancel{7} & \cancel{12} & 17 & \cancel{22} & \cancel{27} \\
\cancel{3} & \cancel{8} & 13 & \cancel{18} & 23 & \cancel{28} \\
\cancel{4} & \cancel{9} & \cancel{14} & 19 & \cancel{24} & 29 \\
\cancel{5} & \cancel{10} & \cancel{15} & \cancel{20} & \cancel{25} & \cancel{30} \\
\cancel{6} & 11 & \cancel{16} & \cancel{21} & \cancel{26} & 31
\end{array}
$$

First, 2 is not a divisor of 1,001; therefore, we eliminate 2 and each multiple of 2. The next number remaining would be 3. By division, 3 is not a divisor of 1,001. We eliminate 3 and each multiple of 3. In the same way, we eliminate 5 and each multiple of 5.

The next number remaining is 7. By division, 7 is found to be a divisor of 1,001; in fact, $7 \times 143 = 1,001$. Conclusion: 1,001 is not a prime number.

By this method, we have obtained the smallest prime divisor of 1,001. By repeating the process with 143 instead of 1,001, we would obtain the next prime divisor of 1,001 (which could possibly be seven again). Continuing in this manner would determine the remaining prime divisors of 1,001, which are 11 and 13. Then, $1,001 = 7 \times 11 \times 13$.

The remaining divisors of 1,001 (other than 1) can be determined by forming all the possible products of the prime divisors 7, 11, and 13:

$$7 \times 11 = 77, \ \ 7 \times 13 = 91, \ \ 11 \times 13 = 143, \ \ 7 \times 11 \times 13 = 1,001.$$

Hence, all of the divisors of 1,001 are

1, 7, 11, 13, 77, 91, 143, and 1,001.

EXAMPLE 1.

Find all of the divisors of 1,386.

The prime divisors of 1,386 are found by continued division as indicated:

2 ⌊1,386
  3 ⌊693
  3 ⌊231
  7 ⌊77
 11 ⌊11
     1

Consequently,

1,386 = 2 × 3 × 3 × 7 × 11.

The possible products of the above factors, two at a time, are:

2 × 3 = 6, 2 × 7 = 14, 2 × 11 = 22, 3 × 3 = 9,
3 × 7 = 21, 3 × 11 = 33, 7 × 11 = 77.

Three at a time:

2 × 3 × 3 = 18, 2 × 3 × 7 = 42, 2 × 3 × 11 = 66,
2 × 7 × 11 = 154, 3 × 3 × 7 = 63, 3 × 3 × 11 = 99,
3 × 7 × 11 = 231.

Four at a time:

2 × 3 × 3 × 7 = 126, 2 × 3 × 3 × 11 = 198,
3 × 3 × 7 × 11 = 693, 2 × 3 × 7 × 11 = 462.

Five at a time:

2 × 3 × 3 × 7 × 11 = 1,386.

The above products together with 1, 2, 3, 7, and 11 are all of the divisors of 1,386. That is, the set of all of the divisors of 1,386 is {1, 2, 3, 6, 7, 9, 11, 14, 18, 21, 22, 33, 42, 63, 66, 77, 99, 126, 154, 198, 231, 462, 693, 1,386}.

In the above example, the division was performed only by prime factors. At each step, these prime factors were found by inspection. If at any step this is not possible, then the sieve method can be used to continue the division. In each case, the final quotient should be 1.

EXAMPLE 2.

Find all the divisors of 1,595.

We first attempt to find all the prime divisors as in Example 1.

$$5\,\underline{|\,1,595}$$
$$319$$

To determine if 319 has a prime divisor, except perhaps itself, we use the sieve and find that $17^2 = 289$ and $18^2 = 324$. Consequently, we look for a possible factor between 2 and 17.

$$
\begin{array}{cccc}
\cancel{2} & \cancel{6} & \cancel{10} & \cancel{14} \\
\cancel{3} & \cancel{7} & \boxed{11} & \cancel{15} \\
\cancel{4} & \cancel{8} & \cancel{12} & \cancel{16} \\
\cancel{5} & \cancel{9} & 13 & 17
\end{array}
$$

The desired factor is 11. Continuing the above division

$$5\,\underline{|\,1,595}$$
$$11\,\underline{|\,319}$$
$$29\,\underline{|\,29}$$
$$1$$

we get $1,595 = 5 \times 11 \times 29$.

The possible products of these factors, two at a time, are:

$$5 \times 11 = 55, \quad 5 \times 29 = 145, \quad 11 \times 29 = 319.$$

Three at a time:

$$5 \times 11 \times 29 = 1,595.$$

The set of all possible divisors is then

$$\{1, 5, 11, 29, 55, 145, 319, 1,595\}.$$

EXERCISES

1. Determine if each of the following numbers is a prime number. Use the Sieve of Eratosthenes only as a last resort.

   (a)† 95   (b) 41   (c)† 37
   (d) 42   (e)† 100   (f) 267
   (g)† 1,062   (h) 53   (i)† 131
   (j) 693   (k)† 1,573   (l) 229
   (m)† 401   (n) 128

2.† Write each number that is not a prime in Exercise 1 as the product of prime numbers.

3.† Find all of the divisors of each number that is not a prime in Exercise 1.

*4.† Complete the given proof of the following theorem in the natural-number system:

> If m and n are natural numbers, and m is not a divisor of n, then no multiple of m is a divisor of n.
>
> *Proof.* Suppose m is not a divisor of n, but some multiple of m, say $m \times k$, is a divisor of n. Then, there is a natural number ℓ, such that $(m \times k) \times \ell = n$. . . .

*5. Complete the given proof of the following theorem in the natural-number system:

> If n is a natural number which is not a prime, then n has a divisor (other than 1) less than or equal to $\sqrt{n}$.
>
> *Proof.* Suppose n is not prime, but has as divisors (other than 1) only numbers greater than the square root of n. Then we can write
>
> $$n = \ell \times m$$
>
> where ℓ and m are each natural numbers greater than $\sqrt{n}$. . . .

3.4 SOME UNSOLVED PROBLEMS IN THE NATURAL NUMBERS

The unsolved or open problems to be considered in this section are statements concerning the natural numbers which have neither been verified to be valid nor proven to be invalid in the natural-number system. These problems are simply stated and easy to understand. However, despite years of work, mathematicians have been unable either to prove or to disprove any of them. The first two to be considered are related to prime numbers.

The first seven even numbers after 2 are 4, 6, 8, 10, 12, 14, and 16. Each of these seven can be written as the sum of two prime numbers:

$$4 = 2 + 2 \qquad 6 = 3 + 3 \qquad 8 = 3 + 5 \qquad 10 = 5 + 5$$
$$12 = 5 + 7 \qquad 14 = 7 + 7 \qquad 16 = 5 + 11$$

Can every even number be so expressed? It seems that it could be done, but this fact has never been proven nor disproven. Consequently, it is an open problem of mathematics, called the *Goldbach conjecture* after the German mathematician Christian Goldbach (1690–1764), who first suggested the possibility.

Formally, the Goldbach conjecture states: every even number greater than 2 can be expressed as the sum of two primes. The conjecture has been verified for even numbers up to 100,000 or more; however, the general statement remains to be proven or disproven.

For a second unsolved problem related to prime numbers, consider the following list of the first twenty-five primes:

$$
\begin{array}{cccc}
2 & \lceil 17 & \lceil 41 & 67 \\
\lceil 3 & \lfloor 19 & \lfloor 43 & \lceil 71 \\
\lfloor 5 & 23 & 47 & \lfloor 73 \\
\lfloor 7 & \lceil 29 & 53 & 79 \\
\lceil 11 & \lfloor 31 & \lceil 59 & 83 \\
\lfloor 13 & 37 & \lfloor 61 & 89 \\
& & & 97.
\end{array}
$$

Indicated in this listing are eight pairs of *twin primes*, that is, prime numbers which differ by 2. Is there a last pair of twin primes, or is there an infinite number?

Just as one might expect that the primes become more scarce among the large natural numbers, one might expect that the gaps between successive primes get to be quite large. Twin primes have been found up to approximately one trillion (1,000,000,000,000), but the general question— do an infinite number of twin primes exist—has never been answered.

Perfect numbers are the subject of our third open problem concerning the natural numbers. A natural number is a *perfect number* if it is the sum of its divisors other than itself.

For example, 6 is a perfect number because

$$6 = 1 + 2 + 3.$$

The smallest perfect number is 6. The next perfect number is 28, for

$$28 = 1 + 2 + 4 + 7 + 14.$$

The third, fourth, and fifth perfect numbers are 496, 8,128 and 33,550,336. There are only 24 known perfect numbers. The largest of these is

$$2^{19,936}(2^{19,937} - 1)$$

which was recently determined to be a perfect number by Bryant Tuckerman at the IBM Research Center.

None of the known perfect numbers are odd, and there is some evidence to suggest that there can be none. But this too has never been proven or disproven. Consequently, the conjecture that there are no odd perfect numbers remains another open problem.

Our final open problem is the question of whether it is possible to find natural numbers x, y, and z that satisfy the equation

$$x^n + y^n = z^n$$

for a positive integer n greater than 2.

When $n = 2$, there are various solutions; for example, $x = 3$, $y = 4$, and $z = 5$ gives

$$3^2 + 4^2 = 5^2,$$

and $x = 6$, $y = 8$, and $z = 10$ gives

$$6^2 + 8^2 = 10^2.$$

This question was first posed by the French mathematician Pierre de Fermat in 1670. He indicated that he had proven that there were no such numbers x, y, and z for all n greater than 2, but he did not give the proof.

Mathematicians have verified that Fermat was correct for $n = 3, 4, 5$, ... up to $n = 25,000$ or more. However, no one has been able to determine if Fermat was correct in his assertion for all n greater than 2. It remains, therefore, an unsolved problem.

There are many more unsolved problems concerning the natural numbers. The four that have been described here are among the more famous. The solution of any of these four would not be of much practical importance, yet each has intrigued mathematicians for many years—and quite likely will continue to do so for many years to come!

EXERCISES

1.† Write each of the even numbers from 18 to 40 as the sum of two prime numbers.

2. Find a pair of twin primes between:
   (a) 100 and 110
   (b)† 130 and 140
   (c) 185 and 195.

3. Verify that 496 is a perfect number.

4.† Verify that 498 is not a perfect number.

5. Verify that 497 is not a perfect number.

3.5 MATH MAGIC

You have probably encountered number tricks in which you are asked to pick a number, do some operations on the number, and, from your final result, are told the number with which you started, your age, or some other secret information. These can be quite amusing, particularly when presented with the proper fanfare.

These tricks are even more interesting for the performer who can not only do the trick, but who also has some insight into why the trick works. Such insight fosters a feeling of superiority—call it smugness, if you like.

We will consider some of these tricks and examine why they work. The tricks all follow a general pattern:

1. Pick a secret number, say n, n usually being a natural number.
2. Perform some given operations on n.
3. Based on the result of the operations, the performer can determine n or some other secret number involved in the operations. Or perhaps he can predict the result with no information at all.

Our format will be to state the series of operations along with an example. Then we will examine why the trick works. In each case, your "prey" will probably need paper and pencil in order to do the operations.

Add an Odd

In this trick, you are able to tell your friend the final result, after he or she has performed the operations, with no knowledge of any of the numbers used.

	Instructions		*Example*
1.	Write down a number greater than zero, but don't tell me what it is.	1.	The number chosen is 147.
2.	Add the next larger number to your choice.	2.	$147 + 148 = 295$
3.	Add 7.	3.	$295 + 7 = 302$
4.	Divide by 2.	4.	$302 \div 2 = 151$
5.	Subtract your original number.	5.	$151 - 147 = 4$
6.	Your answer is 4!		

The answer will always be 4, and you can surprise your friend by announcing sight unseen what the final answer is.

Why it works:

1. Denote the number chosen by n
2. Add the next larger number.

$$\frac{n + 1}{2n + 1}$$

3. Add 7.

$$\frac{7}{2n + 8}$$

4. Divide by 2.

$$\frac{\div 2}{n + 4}$$

5. Subtract the original number.

$$\frac{^-n}{4}$$

The 1 from the next larger number combined with the 7 give a total of 8. Division by 2 then gives the predictable result of 4. The operations with the number n, while disguising the operations on 7 and 1, cancel each other for a net result of 0.

Actually, the number 7 can be replaced by any odd number m and the result will be one-half of $m + 1$. For example,

if 11 is added, the result is $\frac{1}{2}(12) = 6$;

if 17 is added, the result is $\frac{1}{2}(18) = 9$;

if 31 is added, the result is 16; and

if 47 is added, the result is 24.

This allows you to vary the trick if called upon for a repeat performance. In the following illustration, 31 is added.

Instructions	Example
1. Pick a number.	1. 43
2. Add the next larger number.	2. $43 + 44 = 87$
3. Add 31.	3. $87 + 31 = 118$
4. Divide by 2.	4. $118 \div 2 = 59$
5. Subtract your original number.	5. $59 - 43 = 16$
6. Your result is 16!	

Ignoring Digits

In this trick, you are able to tell your friend his original number after he tells you the final result.

Instructions	Example
1. Write down a number greater than zero, but don't tell me the number.	1. Suppose the number is 117.
2. Multiply it by 5.	2. $117 \times 5 = 585$
3. Add 7.	3. $585 + 7 = 592$
4. Multiply by 4.	4. $592 \times 4 = 2,368$
5. Add 8.	5. $2,368 + 8 = 2,376$
6. Multiply by 5.	6. $2,376 \times 5 = 11,880$
7. Tell me your result.	

Ignoring the last two digits, 118$\cancel{80}$, the original number is 1 less than the remaining digits.

Why it works:

1.	Denote the original number by	n
2.	Multiply by 5.	$5n$
3.	Add 7.	$5n + 7$
4.	Multiply by 4.	$\times\ 4$
		$\overline{20n + 28}$
5.	Add 8.	8
		$\overline{20n + 36}$
6.	Multiply by 5.	$\times\ \ 5$
		$\overline{100n + 180}$

Because $100n$ always ends in "00," it will never affect the last two digits in the final result. (In the above example, the $100n$ would be 11,700.) This multiplying by 100 is the heart of the trick. The other operations are mostly to hide the trick.

The number 180 is the product

$$[(7 \times 4) + 8]\,5.$$

The final two digits, 80, are ignored and the 1 subtracted from the remaining digits of the final result.

Because this part of the trick is mostly to mislead, the 7 and 8 can be replaced by any other numbers as long as the product replacing 180 is between 100 and 200. Possible replacements for 7 and 8 are:

6 and 9 (product: $[(6 \times 4) + 9]\,5 = 165$); or,
7 and 6 (product: $[(7 \times 4) + 6]\,5 = 170$).

These replacements allow you to vary the trick from one performance to the next. We conclude with an illustration using 6 and 9.

	Instructions			*Example*
1.	Pick a number.		1.	101
2.	Multiply by 5.		2.	$101 \times 5 = 505$
3.	Add 6.		3.	$505 + 6 = 511$
4.	Multiply by 4.		4.	$511 \times 4 = 2,044$
5.	Add 9.		5.	$2,044 + 9 = 2,053$
6.	Multiply by 5.		6.	$2,053 \times 5 = 10,265$
7.	What is your result?			

To obtain the original number:

$$
\begin{array}{r}
10,2\cancel{65} \\
-\ 1 \\
\hline
101
\end{array}
$$

For added variation, you can ask your friend in Step 1 to write the last two digits of the year in which he or she was born. When the trick is finished, don't announce the original number. Rather, subtract it from the current year. The difference is your friend's age at the end of the current year. In some cases, it may be advisable to tell your friend in advance that you will be able to determine his or her age!

Social Security Mystery

With this trick you might be able to convince your friends that there is hidden meaning in the manner in which social security numbers are assigned.

The basis for this trick is the fact that if one chooses a natural number, scrambles the digits, and subtracts the smaller from the larger, the difference is always a multiple of 9. Suppose, for example, the digits of 356 are scrambled to 563. The difference would be

$$\begin{array}{r} 563 \\ -356 \\ \hline 207 \end{array}$$

and $207 \div 9 = 23$.

Or, scramble 7469 to 4697. The difference is

$$\begin{array}{r} 7469 \\ -4697 \\ \hline 2772 \end{array}$$

and $2772 \div 9 = 308$.

To see why the first difference is divisible by 9, we rewrite the numbers in a different manner.

$$356 = \begin{cases} 300 \\ + 50 \\ + 6 \end{cases} = \begin{cases} 3 \times 100 \\ + 5 \times 10 \\ + 6 \end{cases} = \begin{cases} 3 \times (99 + 1) \\ + 5 \times (9 + 1) \\ + 6 \end{cases}$$

$$= \begin{cases} (3 \times 99) + 3 \\ + (5 \times 9) + 5 \\ + 6 \end{cases}$$
$$\overline{(3 \times 99) + (5 \times 9) + 14}$$

$$563 = \begin{cases} 500 \\ + 60 \\ + 3 \end{cases} = \begin{cases} 5 \times 100 \\ + 6 \times 10 \\ + 3 \end{cases} = \begin{cases} 5 \times (99 + 1) \\ + 6 \times (9 + 1) \\ + 3 \end{cases}$$

$$= \begin{cases} (5 \times 99) + 5 \\ + (6 \times 9) + 6 \\ + 3 \end{cases}$$
$$\overline{(5 \times 99) + (6 \times 9) + 14}$$

Both numbers, 356 and 563, contain the same digits, so the sum of the digits (as indicated in the very last column in each case) will be the same, in this case 14. All of the other numbers have a factor of 9.

In the difference $563 - 356$, which can now be written

$$
\begin{array}{r}
(5 \times 99) + (6 \times 9) + 14 \\
- (3 \times 99) + (5 \times 9) + 14 \\
\hline
\text{- - - - - - - - - - -}\quad 0
\end{array}
$$

the 14's give a net result of 0. Because all of the other numbers have a factor of 9, so will their difference.

In our second illustration above, we have

$$
7469 = \left\{
\begin{array}{lr}
(7 \times 999) & + \ 7 \\
+ (4 \times \ \ 99) & + \ 4 \\
+ (6 \times \ \ \ \ 9) & + \ 6 \\
+ & 9 \\
\hline
\end{array}
\right.
$$
$$(7 \times 999) + (4 \times \ \ 99) + (6 \times 9) + 26$$

$$
4697 = \left\{
\begin{array}{lr}
(4 \times 999) & + \ 4 \\
+ (6 \times \ \ 99) & + \ 6 \\
+ (9 \times \ \ \ \ 9) & + \ 9 \\
+ & 7 \\
\hline
\end{array}
\right.
$$
$$(4 \times 999) + (6 \times \ \ 99) + (9 \times 9) + 26$$

In the difference, $7469 - 4697$, the 26's give a net result of 0; the rest of the numbers all have a factor of 9.

Instructions	Example
1. Write down the last four digits of your social security number.	1. 4263
2. Scramble the digits in any manner you like.	2. 3426

3. Subtract the smaller from the larger of these two numbers.	3. 4263
	    − 3426
	    ‾‾‾‾
	     837

4. If your result is 10 or greater, repeat Steps 2 and 3 until the result is less than 10.	4. (a) 783
	   (b) 837
	      783
	     ‾‾‾‾
	      54 (still greater
	         than 10)
	   (c) 45
	   (d) 54
	      − 45
	     ‾‾‾‾
	       9 (less than 10)

5. Add 2 to this result.	5. 9 + 2 = 11

You can then announce, without seeing any of the calculations, that the final result is either 11 or 2!

Why it works:

Each of the differences in Steps 3 and 4 are divisible by 9. If the process is continued until a difference of one digit ("less than 10") is obtained, that difference must be 9 or 0. The 2 added in Step 5 will always give a final result of 11 or 2.

Actually, the occurrence of 0 for a final result of 2 is quite rare. You can safely announce that the final result is 11, if you are prepared to change your decision to 2 should your friend disagree.

Adding 2 in Step 5 is simply to hide the trick again. The 2 can be replaced by any other number. The final result will then be 9 (or possibly 0) plus this other number. The next illustration adds 121 instead of 2.

	Instructions		Example
1.	Write down the last three digits of your house number in your address.	1.	657
2.	Scramble the digits in any manner.	2.	756
3.	Subtract the smaller from the larger of these two numbers.	3.	756 − 657 = 99
4.	If your result is ten or greater, repeat Steps 2 and 3 until the result is less than 10.	4.	99 − 99 = 0
5.	Add 121 to your result.	5.	121 + 0 = 121

You announce the result to be 130 (that is, 121 + 9). When your friend disagrees, then change to 121. This need to change answers would be very rare.

The process of reducing the difference to a number less than 10 can be quite lengthy. The number of operations can be reduced by asking for a beginning number of only three digits—as in this last illustration.

A Perfect Square

The examination of why this trick works is left for you in the exercises. As with the ignoring digits trick, you will be able to tell your friend the number he or she chose at the beginning.

	Instructions		Example
1.	Pick a number between 0 and 12.	1.	8
2.	Add 2.	2.	8 + 2 = 10
3.	Multiply the two numbers.	3.	8 × 10 = 80
4.	Add 7.	4.	80 + 7 = 87
5.	Tell me your result.		

Mentally subtract 6 from 87 to get 81, find that the square root of 81 is 9, and subtract 1 to proclaim that the original number was 8.

Actually, any number greater than 0 can be used in Step 1. However, the process of mentally taking the square root can get out of hand if the number is too large.

EXERCISES

1.† Examine why the perfect square trick works. (*Hint:* Denote the original number by n, and note that $n^2 + 2n + 1 = (n + 1)^2$.)

2. How can one change the 7 to be added in Step 4 for the sake of variety?

3.6 REPRESENTATION OF THE NATURAL NUMBERS

Imagine that you are adding a column of numbers, one at a time, on a mechanical desk calculator. Although you are only interested in the final sum, the machine has to store each sum in the process in order that the next number you enter can be added to it. This storage is usually performed by a series of counters, perhaps of the type illustrated in Figure 3.2. This counter has ten positions, one for each of the ten digits. It is able to store any one of the ten digits by rotating that digit into the storage position. For example, the counter in Figure 3.3 is storing the digit 7.

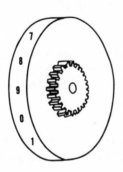
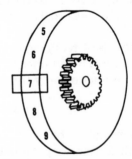

FIGURE 3.2   FIGURE 3.3

Larger numbers are stored by means of a series of interconnected counters. In an 8-place machine, the number 40,169,332 would be stored mechanically by the series of counters shown in Figure 3.4.

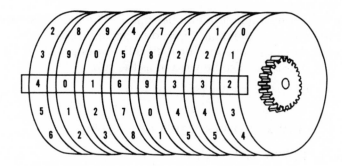

FIGURE 3.4

The counters are rotated by the mechanism of the machine according to the numbers given it and to the operations to be performed. The mechanism by which this is done is not our immediate concern. What does concern us is the fact that each counter can be set in any one of ten positions. Consequently, each counter can store any one of the ten digits.

Electronic computers must also be able to store numbers in some manner. In fact, one of their most useful features is their ability to store vast quantities of numbers.

The basic storage unit (corresponding to the counter above) in many of the large computers is called a *magnetic core*. The magnetic core is a very small ring, made of a material that is easily magnetized, through which a wire is passed. When electrical current is passed through the wire, the magnetic field in the core is polarized in one direction. When the direction of the current is reversed, the magnetic field in the core is likewise reversed (Fig. 3.5). Therefore, the core can be in only two different states or "positions," which correspond to the directions of its magnetic field.

FIGURE 3.5

Our concern is not the electrical process involved in changing the direction of the field. Our concern is the fact that the core can be in only two different states. How, then, can a computer store and manipulate numbers that we usually represent by means of ten different digits? Understanding the answer to this question is our objective in this and the following section.

You might have guessed by now that numbers can be represented by less than ten digits. Our immediate task is to see how this is done.

Consider, for example, the number 5,289. The digit 5 is said to occupy the *thousands* position, the digit 2 is said to occupy the *hundreds* position, the digit 8 the *tens* position, and the digit 9 the *ones* position. The reason for this is that 5,289 can also be written as the sum

$$(5 \times 1{,}000) + (2 \times 100) + (8 \times 10) + (9 \times 1)$$

or

$$(5 \times 10^3) + (2 \times 10^2) + (8 \times 10^1) + (9 \times 10^0).$$

Similarly 43,796 can be written

$$(4 \times 10^4) + (3 \times 10^3) + (7 \times 10^2) + (9 \times 10^1) + (6 \times 10^0).$$

Accordingly, the usual representation of numbers is said to be a *decimal* system, or to be in *base ten*. We could just as well use powers of any other natural number. Suppose we want to use powers of seven. How can we express 5,289 in powers of seven?

The first few powers of seven are

$$7^0 = 1$$
$$7^1 = 7$$
$$7^2 = 49$$
$$7^3 = 343$$
$$7^4 = 2{,}401$$
$$7^5 = 16{,}807$$
$$7^6 = 117{,}649.$$

The highest power of seven less than 5,289 is $7^4 = 2{,}401$. By division

$$
\begin{array}{r}
2 \\
2{,}401\,\overline{)\,5{,}289} \\
\underline{4{,}802} \\
487
\end{array}
$$

we see that

$$5{,}289 = (2 \times 7^4) + 487.$$

We then divide the remainder 487 by the next lower power of seven, which is $7^3 = 343$.

$$
\begin{array}{r}
1 \\
343\,\overline{)\,487} \\
\underline{343} \\
144
\end{array}
$$

Consequently, $487 = (1 \times 7^3) + 144$. Replacing 487 by this expression in the last representation of 5,289 above, we get

$$5{,}289 = (2 \times 7^4) + (1 \times 7^3) + 144.$$

Dividing the remainder 144 by $7^2 = 49$,

$$\begin{array}{r} 2 \\ 49\overline{\smash{\big)}\,144} \\ \underline{98} \\ 46 \end{array}$$

we get

$$5{,}289 = (2 \times 7^4) + (1 \times 7^3) + (2 \times 7^2) + 46.$$

Finally, dividing the remainder 46 by 7 gives

$$\begin{array}{r} 6 \\ 7\overline{\smash{\big)}\,46} \\ \underline{42} \\ 4 \end{array}$$

Because $7^0 = 1$, we can then write

$$5{,}289 = (2 \times 7^4) + (1 \times 7^3) + (2 \times 7^2) + (6 \times 7^1) + (4 \times 7^0).$$

The fact that 5,289 has this representation is written more compactly by the equation

$$5{,}289 = 21264_7.$$

The right-hand side of this last expression is read "two, one, two, six, four, base seven." There are no thousands position, hundreds position, or the like in bases other than base ten.

Instead of performing multiple division by successive powers of seven, the same results can be obtained by continued division by seven itself. The remainder (even if it is 0) is recorded at each division as indicated.

$$\begin{array}{l} 7\,\lfloor\underline{5{,}289} \\ \quad 7\,\lfloor\underline{755} + 4 \\ \quad\quad 7\,\lfloor\underline{107} + 6 \\ \quad\quad\quad 7\,\lfloor\underline{15} + 2 \\ \quad\quad\quad\quad 7\,\lfloor\underline{2} + 1 \\ \quad\quad\quad\quad\quad 0 + 2 \end{array}$$

The remainders, in reverse order, give the representation of 5,289 in base seven.

If one wanted to represent 5,289 in powers of eight (or in base eight) the division would be

$$\begin{array}{l} 8\,\lfloor\underline{5{,}289} \\ \quad 8\,\lfloor\underline{661} + 1 \\ \quad\quad 8\,\lfloor\underline{82} + 5 \\ \quad\quad\quad 8\,\lfloor\underline{10} + 2 \\ \quad\quad\quad\quad 8\,\lfloor\underline{1} + 2 \\ \quad\quad\quad\quad\quad 0 + 1 \end{array}$$

Consequently,

$$5,289 = 12251_8.$$

This last expression means

$$5,289 = (1 \times 8^4) + (2 \times 8^3) + (2 \times 8^2) + (5 \times 8^1) + (1 \times 8^0).$$

Sometimes the subscript 10 will be used to emphasize the fact that a particular number is written in base ten. For example,

$$5289_{10} = 21264_7$$

and

$$5289_{10} = 12251_8.$$

EXAMPLE 1.

Write 16,432 in base nine.

$$
\begin{array}{r}
9\,\underline{|\,16,432} \\
9\,\underline{|\,1,825} + 7 \\
9\,\underline{|\,202} + 7 \\
9\,\underline{|\,22} + 4 \\
9\,\underline{|\,2} + 4 \\
0 + 2
\end{array}
$$

Therefore,

$$16432_{10} = 24477_9.$$

EXAMPLE 2.

Write 432 in base two.

$$
\begin{array}{r}
2\,\underline{|\,432} \\
2\,\underline{|\,216} + 0 \\
2\,\underline{|\,108} + 0 \\
2\,\underline{|\,54} + 0 \\
2\,\underline{|\,27} + 0 \\
2\,\underline{|\,13} + 1 \\
2\,\underline{|\,6} + 1 \\
2\,\underline{|\,3} + 0 \\
2\,\underline{|\,1} + 1 \\
0 + 1
\end{array}
$$

Consequently,

$$432 = 110110000_2.$$

EXAMPLE 3.

Write 432_5 in base ten.

$$432_5 = (4 \times 5^2) + (3 \times 5^1) + (2 \times 5^0)$$
$$= (4 \times 25) + (3 \times 5) + (2 \times 1)$$
$$= 100 + 15 + 2$$
$$= 117$$

Consequently,

$$432_5 = 117_{10}$$

or

$$432_5 = 117.$$

Note that the procedure used to convert *to* base ten in this example differs from the procedure used in Examples 1 and 2 to convert *from* base ten.

EXAMPLE 4.

Write 110101_2 in base ten.

$$110101_2 = (1 \times 2^5) + (1 \times 2^4) + (0 \times 2^3) + (1 \times 2^2)$$
$$+ (0 \times 2^1) + (1 \times 2^0)$$
$$= (1 \times 32) + (1 \times 16) + (0 \times 8) + (1 \times 4)$$
$$+ (0 \times 2) + (1 \times 1)$$
$$= 32 + 16 + 0 + 4 + 0 + 1 = 53$$

Therefore,

$$110101_2 = 53.$$

EXAMPLE 5.

Write 504_6 in base three.

We first write 504_6 in base ten and then convert the result to base three.

$$504_6 = (5 \times 6^2) + (0 \times 6^1) + (4 \times 6^0)$$
$$= (5 \times 36) + (0 \times 6) + (4 \times 1)$$
$$= 180 + 0 + 4 = 184$$

Then

```
3 ⌊184
 3 ⌊61 + 1
  3 ⌊20 + 1
   3 ⌊6 + 2
    3 ⌊2 + 0
     0 + 2
```

Therefore,

$$504_6 = 20211_3.$$

EXERCISES

1. Write each of the following numbers in the base indicated.

    (a)† 47 in base two   (b)   98 in base nine
    (c)† 194 in base five   (d)   743 in base six
    (e)† 222 in base two   (f)   397 in base two
    (g)† 1,378 in base eight   (h)   5,246 in base four

2. Write each of the following numbers in base ten.

    (a)† 101_2 (b) 1643_7
    (c)† 222_3 (d) 222_4
    (e)† 11011_2 (f) 4601_8
    (g)† 4001_5 (h) 100011_2

3. Write each of the following numbers in the base indicated.

    (a)† 310_4 in base two   (b) 601_7 in base five
    (c)† 1011_2 in base four   (d) 499_{10} in base six

*4. Suppose that we have invented two new digits, A for 10 and B for 11.

    (a)† Show that $5,036 = 2AB8_{12}$.
    (b)   Write 1,451 in base twelve.

3.7 COMPUTER ARITHMETIC

We have just seen that numbers can be represented in bases other than ten. What makes this fact useful in the construction of computers is that less digits are required as the base becomes smaller. While the representation of numbers in base ten uses all the digits 0 through 9, representation in base seven, for example, uses only the digits 0 through 6. Representation in base four uses only the digits 0 through 3.

In particular, representation in base two requires only the digits 0 and 1. Consequently, any device used to store numbers in base two needs to store only 0's and 1's. This is just what happens in computers.

Recall that a magnetic core, the basic storage unit for computers discussed in Section 3.6, can only be in one of two positions. It stores a 0 when its magnetic field is in one direction and a 1 when the direction is reversed. This correspondence is illustrated in Figure 3.6. To store the number $13 = 1101_2$, for example, would require four magnetic cores. Their magnetic fields would be such that the first two and the fourth would correspond to a 1, and the third to a 0.

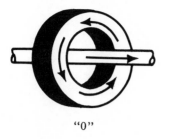

"0"   "1"

FIGURE 3-6

Not only are numbers stored in base two, but arithmetic is also performed by the computer in base two. Arithmetic in base two is called *binary arithmetic*, and numbers written in base two are called *binary numbers*. The remainder of this section is devoted to binary arithmetic. The processes involved are illustrated in the examples. In all of the computations, the base two subscript will be omitted.

EXAMPLE 1.

Add 11011_2 and 1010_2.

```
          1 1 0 1 1
            1 0 1 0
(Carries)       1
          1
        1 0 0 1 0 1
```

$1 + 0 = 1$ Write 1.
$1 + 1 = 2 = 1(2) + 0$ Write 0, carry 1.
$0 + 0 + 1 = 1$ Write 1.
$1 + 1 = 2 = 1(2) + 0$ Write 0, carry 1.
$1 + 1 = 2 = 1(2) + 0$ Write 10.

We can check our result by converting to base ten and then perform the addition in base ten.

Check:

$$1\ 1\ 0\ 1\ 1 = 27_{10}$$
$$+\ \ \ 1\ 0\ 1\ 0 = 10_{10}$$
$$37_{10} = 1\ 0\ 0\ 1\ 0\ 1$$

EXAMPLE 2.

Add 1011_2, 111_2, and 1011_2.

```
           1 0 1 1
             1 1 1
           1 0 1 1
(Carries)      1
           1 0
           1 1 1 0 1
```

$1 + 1 + 1 = 3 = 1(2) + 1$ Write 1, carry 1.
$1 + 1 + 1 + 1 = 4 = 1(2^2) + 0(2) + 0$
                                Write 0, carry 10.
$0 + 1 + 0 + 0 = 1$ Write 1.
$1 + 1 + 1 = 3 = 1(2) + 1$ Write 11.

Check:

$$1\ 0\ 1\ 1 = 11_{10}$$
$$1\ 1\ 1 = 7_{10}$$
$$1\ 0\ 1\ 1 = \underline{11_{10}}$$
$$29_{10} = 1\ 1\ 1\ 0\ 1$$

EXAMPLE 3.

Add 1111_2, 101_2, and 1101_2.

```
            1 1 1 1
              1 0 1
            1 1 0 1
(Carries)       1
              1
          1 0
          1
          1 0 0 0 0 1
```

$1 + 1 + 1 = 3 = 1(2) + 1$ Write 1, carry 1.
$1 + 0 + 0 + 1 = 2 = 1(2) + 0$ Write 0, carry 1.
$1 + 1 + 1 + 1 = 4 = 1(2^2) + 0(2) + 0$
                                Write 0, carry 10.
$1 + 1 + 0 = 2 = 1(2) + 0$ Write 0, carry 1.
$1 + 1 = 2 = 1(2) + 0$ Write 10.

Check:

$$1\ 1\ 1\ 1 = 15_{10}$$
$$1\ 0\ 1 = 5_{10}$$
$$1\ 1\ 0\ 1 = \underline{13_{10}}$$
$$33_{10} = 1\ 0\ 0\ 0\ 0\ 1$$

After you are more familiar with the procedure, addition can be written in the usual manner, without so much space devoted to the carries.

Subtraction is fairly routine in base two, as Example 4 illustrates. The only complication enters when a "borrow" is involved, in which case it is essential to remember that

$$10_2 = 1(2^1) + 0(2^0) = 2 + 0 = 2.$$

Examples 5 and 6 illustrate the procedure when a "borrow" is required.

EXAMPLE 4.

Subtract 100_2 from 1101_2.

$$\begin{array}{r} 1101 \\ -\ 100 \\ \hline 1001 \end{array}$$

Check:

$$\begin{array}{r} 1101 = 13_{10} \\ -\ 100 = \underline{\ 4_{10}} \\ 9_{10} = 1001 \end{array}$$

EXAMPLE 5.

Subtract 10_2 from 1101_2.

$$\begin{array}{r} 0 \\ 1\ 1^10\ 1 \\ -\quad 1\ 0 \\ \hline 1\ 0\ 1\ 1 \end{array}$$

$10 - 1 = [1(2) + 0] - 1 = 2 - 1 = 1$

Check:

$$\begin{array}{r} 1101 = 13_{10} \\ -\quad 10 = \underline{\ 2_{10}} \\ 11_{10} = 1011 \end{array}$$

EXAMPLE 6.

Subtract 1011_2 from 11110_2.

$$\begin{array}{r} 0^10 \\ 1\ 1\ 1\ 1^10 \\ -\quad 1\ 0\ 1\ 1 \\ \hline 1\ 0\ 0\ 1\ 1 \end{array}$$

$10_2 - 1 = 1$

EXERCISES

1.† Write each of the numbers from 1 to 20 in base two.

2. Write each of the following base two numbers in base ten.

(a) 11   (b) 100   (c) 101
(d) 111   (e) 1001   (f) 1101
(g) 10010   (h)† 11010   (i)† 101010
(j)† 111111

3. Perform the following additions in base two. Check each of the first four by converting to base ten.

(a)†   10   (b)   11   (c)†   101
       101   110   1011

(d)   10   (e)†   11   (f)   10
       101   100   111
       111   1010   1000

(g)†   10   (h)   111
       1011   1111
       1111   11111

4. Perform each of the following subtractions in base two. Check each of the first four by converting to base ten.

(a)† 1101   (b)   1011   (c)† 1010
     − 101   − 101   − 101

(d)   100   (e)† 1110   (f)   10001
     −   1   − 111   − 1101

(g)† 1000   (h)   100010
     − 111   − 11101

Multiplication and division in base two are also fairly routine. However, one must remember that the addition involved in multiplication is binary addition. Likewise, the subtraction in division is binary subtraction. The remaining examples illustrate binary multiplication and binary division.

EXAMPLE 7.

Multiply 1011_2 and 101_2 in base two.

$$
\begin{array}{r}
1\,0\,1\,1 \\
\times\ \ 1\,0\,1 \\
\hline
1\,0\,1\,1 \\
0\,0\,0\,0 \\
1\,0_{,1}1 \\
\hline
1\,1\,0\,1\,1\,1
\end{array}
$$

Check:

$$
\begin{array}{r}
1011 = 11_{10} \\
\times\ \ 101 = \ \ 5_{10} \\
\hline
55_{10} = 110111
\end{array}
$$

EXAMPLE 8.

Multiply 10011_2 and 1101_2 in base two.

$$
\begin{array}{r}
1\,0\,0\,1\,1 \\
\times\ \ 1\,1\,0\,1 \\
\hline
1\,0\,0\,1\,1 \\
0\,0\,0\,0\,0 \\
1\,0\,0\,1\,1 \\
1\,0\,0_{,1,1}1 \\
\hline
1\,1\,1\,1\,0\,1\,1\,1
\end{array}
$$

EXAMPLE 9.

Divide 110111_2 by 101_2 in base two.

$$
\begin{array}{r}
1\,0\,1\,1 \\
101\overline{)1\,1\,0\,1\,1\,1} \\
\underline{1\,0\,1} \\
1\,1 \\
\underline{0} \\
1\,1\,1 \\
\underline{1\,0\,1} \\
1\,0\,1 \\
\underline{1\,0\,1} \\
0
\end{array}
$$

Check: Instead of converting, we can compare this result with Example 7, where we found

$$101_2 \times 1011_2 = 110111_2.$$

EXAMPLE 10.

Divide 11110111_2 by 1101_2 in base two.

```
              1 0 0 1 1
      1 1 0 1 ⟌ 1 1 1 1 0 1 1 1
              1 1 0 1
                1 0 0
                    0
              1 0 0 1
                    0
              1 0 0 1 1
                1 1 0 1
                1 1 0 1
                1 1 0 1
                      0
```

Check: Compare with Example 8.

The binary arithmetic operations just examined are the same as those performed by the computer. However, the computer has no mental facility for performing the operation as you do. Rather, the operations are performed electronically by the circuits in the computer's processing unit. The speed at which the operations are performed is measured in nanoseconds, or billionths of a second. Try to imagine how long one nanosecond must be!

EXERCISES

1. Perform each of the following multiplications in base two. Check the result of the first four by converting to base ten.

   (a)†   1010   (b)   1001   (c)†   111
        X 101   X 111   X 11

   (d)   1011   (e)†   1011   (f)   11011
        X 11   X 101   X 110

   (g)†   10111   (h)   11011
        X 101   X 1011

2. Perform each of the following divisions in base two. Check the result of the first four by comparing with the corresponding problem in Exercise 1.

   (a)† $110010 \div 101$ (b) $111111 \div 111$
   (c)† $10101 \div 11$ (d) $100001 \div 11$

(e)† 100011 ÷ 101   (f) 100011 ÷ 111
(g)† 1000010 ÷ 110   (h) 11100111 ÷ 1011

3.8 SUMMARY

The mathematical system concept extends to the mathematics most familiar to most people—arithmetic, geometry, and algebra. Euclidean geometry
is itself a mathematical system; arithmetic and algebra are performed in
the real-number system. The development of the real-number system begins
with the development of the natural numbers. The development of the
natural numbers, in turn, is based on only four axioms and the definitions
of addition and multiplication. Properties of the natural numbers used in
calculations (for example, the Commutative Laws of Addition and Multiplication) then follow as theorems.

Prime numbers make up a special class of natural numbers, a prime
number being a natural number having no divisors other than itself and 1.
By forming all possible products of the prime divisors of a given natural
number, it is possible to determine all of the divisors of the given number.
We must be able to determine all of the divisors of a natural number to
decide if the number is a perfect number—a perfect number being a natural
number that is equal to the sum of all of its divisors other than itself. The
study of prime numbers and of perfect numbers has led to a number of
famous unsolved problems in mathematics.

Two special topics considered relative to the natural numbers were
number games and computer arithmetic. Although the games can be performed with no insight into how they work, such insight can provide
added enjoyment.

The usual representation of natural numbers is in base ten, using the
ten digits 0, 1, 2, 3, 4, 5, 6, 7, 8, and 9. However, they can be represented
just as well in other bases, using less than ten digits when the base is less
than ten. In particular, only the two digits 0 and 1 are required in base
two representation. The corresponding arithmetic using only the 0 and 1
digits is called binary arithmetic. This is the arithmetic performed by computers. The results are usually converted to base ten when they are printed.
However, this conversion is primarily a function of the printing device;
inside the computer, the numbers are all in terms of 0's and 1's.

"Mathematics is the gate and key of the sciences .... Neglect of mathematics works injury to all knowledge, since he who is ignorant of it cannot know the other sciences or the things of this world."

ROGER BACON (1214?–1294)

M. C. Escher: *Relativity*, 1953. An interesting effect obtained by unusual use of perspective is characteristic of much of Escher's work. (See Section 4.3.)

chapter 4

mathematical models

In Chapter 1, it was pointed out that every application of mathematics in solving a physical problem involves two basic features—a mathematical system and a mathematical description. Both of these features were used in the previous chapters, with more attention devoted to the system aspect and less to the mathematical description concept. The purpose of this chapter is to examine more fully this latter concept and its relationship to mathematical systems.

4.1 EXAMPLES AND USE OF MATHEMATICAL MODELS

A *mathematical model* is a mathematical description of a physical situation. Here the term "physical" is used in the very broad sense of including any real-life situation, be it financial, biological, administrative, sociological, or whatever. The mathematical model is the means by which the transition is made from the physical world to the mathematical. Every application of mathematics requires that this transition be made; consequently, every application of mathematics requires a mathematical model.

For the sake of illustration, examine the process by which Euler solved the Königsberg bridge problem described in Chapter 1.

1. There was a physical problem—determine if a person could take a walk in such a way as to cross each of the bridges exactly once.
2. The graph in Figure 4.1 was constructed to describe the situation mathematically; that is, a mathematical model was constructed. The physical problem then became the mathematical problem of determining if the graph had an Euler path.

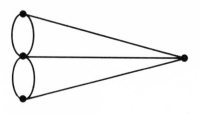

FIGURE 4.1

3. The solution of the mathematical problem was found in the corresponding mathematical system (graph theory)—no Euler path was possible because the graph had more than two odd nodes.
4. The mathematical solution was interpreted in the physical situation—the single crossing of all the bridges was not possible.

Note that the transition to the mathematical world occurred in the second step with the construction of the mathematical model. The transition back to the physical world occurred in the fourth step in the interpretation of the solution.

Each application of probability in Chapter 2 involved a similar four-step procedure. For example, consider the solution of the question, "What is the likelihood of obtaining a sum of 5 in a roll of a pair of dice?"

1. A question is asked concerning a physical situation.
2. A mathematical model is constructed—the sample space with subsets as events and the probability function, which assigns a value of $\frac{1}{36}$ to each element in the sample space. The original question becomes the mathematical question, "What is the probability of the event $E = \{(1,4), (4,1), (2,3), (3,2)\}$?"
3. A mathematical solution is found in the system of probability— $P(E) = \frac{1}{9}$.
4. The mathematical solution is interpreted in the physical situation—in the long run, a sum of 5 will be obtained on the roll of a pair of dice $\frac{1}{9}$ of the time, and in this sense indicates the degree of certainty of obtaining a sum of 5 on any future roll.

Each of the above illustrations follows the same general pattern. Each involves a physical situation with a relevant question. An appropriate mathematical model is chosen to describe the physical situation, and the question is translated into mathematical terms. A solution of the question, as mathematically formulated, is found in a corresponding mathematical system. Then the solution is translated back into physical terms. This procedure is outlined in Figure 4.2.

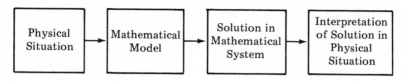

FIGURE 4.2

If an appropriate mathematical model is not readily available, then a great amount of experimental work and testing of possible models (not reflected in Figure 4.2) is usually involved. The testing of a possible model amounts to determining if the solution given by the model to the physical problem agrees with the result of the experimental work. If the solution is not in harmony with experimental results, the model must be discarded, and a new model sought.

Suppose, for example, that you are a social scientist and are interested in predicting the annual income of an individual on the basis of the number of years of education and the number of years of employment. You collect this information on 1,000 individuals with varying educational and employment backgrounds. You then construct a mathematical model to relate annual income to years of education and years of employment and test the model with your data. If the model gives the appropriate annual income (with some reasonable degree of accuracy) for only 225 of the individuals that you studied, it wouldn't be considered very reliable in predicting annual income. If you're persistent, you then attempt to construct another model. You may or may not be successful in the end.

The advances of physics and engineering are due, to a great extent, to the availability of good mathematical models. Because the mathematical approach has served the sciences so well, other areas of study have begun to adapt the mathematical method to their own problems. One result has been the establishment of new branches of study within these other areas, such as econometrics in economics, biomathematics in biology, and operations research in business administration.

You probably have encountered the use of mathematical models many times, perhaps several times in the course of a single day, without recognizing the process in terms of models. The next several examples illustrate the use of mathematical models. The first four occur in rather technical circumstances. Do not be concerned about the technical details. These examples are intended to illustrate the variety of ways in which mathematical models are used. The technical details are not our concern.

EXAMPLE 1.

The mathematical models used by economists for determining the gross national product consist of up to 400 equations involving as many as 1,000 variables. These, of course, take years to develop and are much too lengthy to present here. Instead we shall construct our own simplified version, recognizing that it probably is not very reliable.

The gross national product is the total production of goods and services in this country for one year. We shall consider the production of goods and services to be of four types:

1. Consumer spending (C).
2. Investments (I).
3. Government spending (G).
4. Net exports (the difference between exports and imports) (X_n).

Then the gross national product (Y) is given by

$$Y = C + I + G + X_n.$$

In addition, we will assume that consumer spending is approximately $\frac{9}{10}$ of the gross national product, and that some consumer spending would still occur even if the gross national product were 0. This gives

$$C = a + 0.9Y$$

where a is the amount of consumer spending that would occur when the gross national product (Y) is 0.

Our model then consists of the two equations

$$Y = C + I + G + X_n$$
$$C = a + 0.9Y.$$

If we substitute the value of C from the second equation into the first we get

$$Y = (a + 0.9Y) + I + G + X_n$$

which gives

$$Y - 0.9Y = a + I + G + X_n.$$

Combining the two terms on the left gives

$$\tfrac{1}{10} Y = a + I + G + X_n$$

or

$$Y = 10(a + I + G + X_n).$$

According to our model, an increase of one dollar in government spending (G is increased by 1), for example, would generate an additional ten dollars in the gross national product.

It should be noted that our model is not entirely unrealistic, as the equation

$$Y = C + I + G + X_n$$

is the basic equation in most models actually used by economists.

EXAMPLE 2.

A certain bacterium reproduces in such a way that there is about 10% more bacteria present at the end of each hour than there was at the beginning. In biology, such bacteria are said to have an *exponential growth rate*. The mathematical model that gives the number present (y) at any time (t) is given by

$$y = ke^{0.1t}$$

where k is the number present when the counting process begins, and e is a constant approximately equal to 2.718. (The number e, like π, is an infinite nonrepeating decimal which can be approximated to as many decimal places as desired.)

If a sample of the above bacterium contains 300 bacteria, how many will it contain ten hours later?

In this case, $t = 10$ and $k = 300$, so that the number present ten hours later would be

$$300e^{(0.1)10} = 300e \simeq 300(2.718) = 815.4.$$

EXAMPLE 3.

When a space ship is launched from Cape Kennedy, its initial velocity (speed) must be sufficient to overcome the effect of the Earth's gravity. How great must the velocity be?

Newton's law of gravity states that the force of gravity on an object is inversely proportional to the square of its distance from the center of the Earth. The mathematical model which expresses the effect of gravity (a) on an object is then

$$a = \frac{k}{r^2}$$

where k is a constant and r is the distance of the object from the center of the Earth.

This equation can be solved, using the methods of calculus, to determine that the minimum initial velocity must be about seven miles per second.

EXAMPLE 4.

An electric circuit is a closed system in that the current passing through the system returns to the source from which it originated. This arrangement is illustrated in Figure 4.3. There is a voltage drop in the circuit as the current passes through the light bulb because of the electricity used by the bulb.

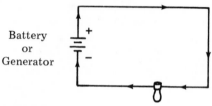

Battery
or
Generator

FIGURE 4.3

Items which use electricity, such as a light bulb or a toaster, are called *resistors*, and individuals who study electric circuits or design appliances are interested in determining the voltage drop due to the current passing through resistors.

Ohm's law, verified experimentally, provides an expression for the voltage drop in terms of the type of resistor and the amount of current from the power source: the voltage drop across a resistor is proportional to the current.

A mathematical model of the voltage drop would then be

$E = RI$

where E denotes voltage drop, I denotes the amount of current, and the constant R depends upon the resistor. The value of R would be small for a light bulb, which uses little electricity, as compared to a toaster, which would require a larger value of R. When two of the three quantities E, R, and I are known, the third can be determined from the above equation.

EXAMPLE 5.

The story problems in an algebra course all entail the use of a mathematical model. A common type of problem of this sort is the following.

If Betty is twice as old as Paul, and in three years Paul will be twenty-one years old, how old is Betty?

The mathematical model of the situation is: let y be Betty's age, and let x be Paul's age. Then

$$y = 2x$$
$$x + 3 = 21.$$

The mathematical question is, "What is the value of y?" The pair of equations is solved in the system of real numbers to determine that the value of y is 36. The physical interpretation is that Betty is 36 years old.

EXAMPLE 6.

A common use of mathematical models is in connection with checking accounts. This usually occurs in the check book, where a running account of the bank balance is kept. This is the mathematical model of the account, and it answers the question, "How much money do I have?" Calculations are made in the system of rational numbers (3.57 for example, is equal to 357/100.) Deposits are accounted for by adding the rational number equal to the amount deposited. Checks written are accounted for by subtracting the rational number equal to the amount of the check written. The balance at any time answers the above question.

In the exercises, you are asked to construct and use a number of mathematical models that involve quantities measured by real numbers. Construction of these models amounts to assigning letters to represent the variable or unknown quantities involved, and then to expressing in a mathematical way the physical relationship between these variables.

EXAMPLE 7.

In a particular congressional district there are 250,000 Democratic voters. The polls indicate that if the Democratic incumbent would run again, he would get the votes of all of the Democrats and 20% of the Republicans.

(a)   Construct a mathematical description of the total number of votes the incumbent would get.

The unknown quantities are the number of Republican voters and the total number of votes for the incumbent. Let x denote the number of Republican voters and y the total number of votes. Then

$$y = 250{,}000 + 0.20x$$

is the desired model.

(b)   If the actual number of Republican voters was 300,000, how many votes would the incumbent receive?

If we set x (the number of Republicans) equal to 300,000 in the above equation, we get

$$y = 250{,}000 + 0.20(300{,}000)$$

or

$$y = 250{,}000 + 60{,}000.$$

He would get 310,000 votes.

In part (a) of Example 7, the verbal expression of the desired relationship is

Total votes equals Democratic votes and 20% of Republican votes.

Note how closely the mathematical expression parallels this verbal expression.

Total votes equals Democratic votes and 20% of Republican votes
y $=$ 250,000 $+$ $0.20x$

This parallel relationship can often be used to construct the mathematical expression. First determine the verbal relationship; then convert the verbal relationship to mathematical form. This procedure is illustrated in Example 8.

EXAMPLE 8.

A secretary estimates that it takes her twelve minutes to type a particular form letter. Write an equation to express the number of letters typed (y) in terms of the total time (x) spent typing the letter.

The number of letters typed is found by dividing the total time by 12, so the verbal relationship is:

Number of letters is equal to total time divided by 12.

This expression can then be converted to mathematical form as follows:

Number of letters is equal to total time divided by 12.
y $=$ x $\div$ 12

or,

$$y = x \div 12.$$

The models which you are asked to construct in the exercises, as well as most of the above examples, are *equations*. However, you should not infer that this must always be the case. In every application of graph theory in Chapter 1, the mathematical model was a *graph*. In every application of probability, the sample space (which was a mathematical description of all possible outcomes) was a *set*. In later chapters, mathematical models take various other forms. At times, they may be nothing more than a number to denote a measurement or an amount.

EXERCISES

The following situations each involve relationships between quantities which are measured in terms of real numbers. Construct mathematical models which express these relationships, using the notation indicated. Then use the models to answer the given questions. (The models constructed here will be used again in Chapter 5.)

1. (a)† The owner of a book store estimates that one-tenth of the books that he puts on the shelves of his store "disappear" due to shoplifters. Write an equation to express the number of books stolen (y) in terms of the number of books put on the shelf (x).

   (b)† During a particular month, the owner shelved 500 new books. How many of these will disappear?

   (c) Which aspects of parts (a) and (b) enter into each step of the four-step procedure (Fig. 4.2) for solving physical problems?

2. (a) The principal of a high school has found that about 7% of each class of entering freshmen become dropouts before their class graduates. Write an equation to express the number of dropouts (y) in terms of the number of entering freshmen (x). *Note: Percent* means per hundred; so for any positive number c, $c\%$ is equal to $c/100 = c \times \frac{1}{100} = c \times 0.01$. For example,

$$9\% = 9 \times 0.01 = 0.09$$
$$12\% = 12 \times 0.01 = 0.12.$$

   (b) This year there were 250 entering freshmen. How many of these will become dropouts?

   (c) Which aspects of parts (a) and (b) enter into each step of the four-step procedure (Fig. 4.2) for solving physical problems?

3. (a)† The quality control manager of an electrical supplies firm estimates that 11% of the light bulbs manufactured by the firm are defective. Write an equation to express the number of defective bulbs (y) in terms of the total number of bulbs manufactured (x).

   (b)† Last week, the company manufactured 10,000 light bulbs. How many of these were defective?

4. (a) The city council of a certain city has authorized one patrolman for each 10,000 residents. Write an equation to express the number of policemen authorized (y) in terms of the number of residents (x).

   (b) How many patrolmen are authorized if there are 763,429 residents? (Round any fraction to the nearest whole number.)

5. (a)† Past experience indicates that 2 out of every 25 bills mailed each month by the credit manager of a certain department store will

not be paid until a reminder notice is sent. Write an equation to express the number of reminders that will have to be sent (y) in terms of the number of bills mailed (x).

(b)† During March, the credit manager sent out 250 reminder notices. How many bills did he send out?

6. (a) In conversion to the metric system of measurement, each yard is equal to 0.90 meters. Write an equation to express the number of meters (y) in any number of yards (x).

(b) A football field is 100 yards long. How many meters is the length of the field?

7. (a)† Herman wants to construct a plywood storage trunk. He estimates his expenses to be a total of $3.00 for hardward and 50¢ for each square foot of plywood. Write an equation to express the total cost (y) of the trunk in terms of the number of square feet (x) of plywood used.

(b)† If Herman decides to spend $15.00 on the trunk, how much plywood should he buy?

8. (a) In anticipation of his income tax return, Jim decides that he will save $200 plus $\frac{1}{10}$ of the total amount he receives from his return. Write an equation to express the total amount that he will save (y) in terms of the amount he receives (x).

(b) If Jim receives a return of $450.50, how much will he save?

9. (a)† A basketball player signs a contract that pays him $25,000 plus $1,000 for each game the team wins. Write an equation to express his total salary (y) in terms of the number of games won (x).

(b)† If his total salary for the year is $51,000, how many games did the team win?

10. (a) As a television salesman, Tom makes a monthly salary of $200 plus a commission of $50 for each television set he sells. Write an equation to express Tom's total salary (y) in terms of the number of television sets that he sells (x).

(b) During April, Tom sold 17 television sets. What was his salary that month?

11.† By analyzing the population behavior of a certain country over a number of years, sociologists have determined that the number of people living in that country at the end of a year is equal to the number of immigrants plus 110% of the number of people living in the country at the beginning of the year. If the number of immigrants during a particular year is 2,000, write an equation to express the total population of the country (y) at the end of that year in terms of the population of the country (x) at the beginning of the year.

12. Dana manufactures and sells a modernistic chair that she has designed. Her profit per chair is $7.00, not taking into account her annual overhead expenses of $2,300. Write an equation to express her annual profit (y) in terms of the number of chairs sold during the year (x).

13.† An airline company has scheduled a 100-seat airplane for a charter flight to the Super Bowl. The price per person is $140 plus $5.00 additional for each empty seat.

   (a) Write an equation to express the number of people (w) on the flight in terms of the number of empty seats (x).
   (b) Write an equation to express the price per person (z) in terms of the number of empty seats (x).
   (c) Using parts (a) and (b) above, write an equation to express the income of the airline company (y) from the flight in terms of the number of empty seats (x).
   (d) If there are 83 people on the flight, how much did each passenger pay?

14. A farmer has an apple orchard with 90 trees. He estimates that his annual profit per tree is $10.00 and that, for each additional tree that he plants in the orchard, his profit per tree will decrease by 10¢.

   (a) Write an equation to express the total number of trees (w) in terms of the number of additional trees planted (x).
   (b) Write an equation to express the profit per tree (z) in terms of the number of additional trees planted (x).
   (c) Using parts (a) and (b), write an equation to express the farmer's total profit (y) in terms of the number of additional trees planted.
   (d) If he plants 20 additional trees, what is his total profit?

15.† Pat intends to convert a rectangular part of his backyard into a garden and has 60 feet of fencing that he intends to use to surround the plot. At first he plans to allow 15 feet of fencing per side, but then suspects that as he varies the length of the sides he might possibly obtain a larger area.

   (a) Denote the length of the garden plot by x. Write an equation to express the width (w) in terms of x, keeping in mind that the plot must be able to be surrounded by 60 feet of fencing.
   (b) Using part (a), write an equation to express the area (y) of the plot in terms of the length (x). *Note:* Area of a rectangle is length times width.
   (c) What is the area of the plot if the width is 10 feet?

16. Paul owns an apartment complex with 20 units. He estimates that he can rent all 20 units if he charges $150 per unit per month and that, for each increase of $25 in rent, one apartment will be vacated.

   (a)   Considering each vacant apartment to correspond to a $25 increase in rent, write an equation to express the rent per apartment (r) in terms of the number of empty apartments (x).

   (b)   Write an equation to express the number of apartments occupied (z) in terms of the number of apartments empty (x).

   (c)   Using parts (a) and (b), write an equation to express the total income per month (y) in terms of the number of empty apartments (x).

   (d)   If Paul has four empty apartments, what rent is he charging?

4.2   ERRORS IN THE USE OF MATHEMATICAL MODELS

Errors in the use of mathematical models can occur at each step of Figure 4.2. In connection with the choice of a mathematical model, the question to be answered must be taken into consideration; a model appropriate for the information desired must be chosen. This is usually the most difficult aspect of applied mathematics.

An occurrence of this type of error—the choosing of an inappropriate model—would be in connection with Example 5 of Section 4.1 if the set of equations were chosen to be

$$y = 2x$$
$$x - 3 = 21,$$

even though this sytem of equations was then solved correctly to obtain Betty's age as 48 and Paul's age as 24.

Even though an appropriate model may be chosen, a second type of error can occur by faulty calculations within the corresponding mathematical system. An example of this type of error would be the situation in which the system of equations in Example 5

$$y = 2x$$
$$x + 3 = 21$$

was solved incorrectly to obtain $x = 7$ and $y = 14$. Here the correct mathematical description was chosen; the error was in the calculations.

A third type of error occurs when the correct model is chosen, and the correct calculations are made, but the results of the calculations are given an incorrect interpretation in the physical situation. An easy example of this could occur in Example 5 if the system of equations

$$y = 2x$$
$$x + 3 = 21$$

was correctly solved to obtain $x = 18$ and $y = 36$, but the conclusion was made that Betty is 18 and Paul is 36.

Or consider again the bank account in Example 6 of Section 4.1. Suppose that correct calculations result in a final total of -13.26. An error in interpretation of the results would occur if a person interpreted this to mean that there was $13.26 in the account, rather than a deficit of that amount!

EXERCISES

In each of the exercises below, a physical problem is solved mathematically. Each instance involves an error in the use of mathematical models. Find the error and identify the type of error as (a) an incorrect model, (b) faulty operations in the real number system, or (c) erroneous interpretation of the mathematical solution.

1.† A savings and loan association pays 5% interest per year on deposits. Betty want to deposit a sum of money which will give her $200 interest per year. How much should she deposit?

Erroneous Solution: If x denotes the amount of money deposited and y the interest received, the *mathematical model* that expresses the relationship between x and y is given by

$$y = 0.05x.$$

Because $200 interest is desired, let y be 200 and *calculate* the corresponding value of x:

$$200 = 0.05x$$
$$\frac{200}{0.05} = x$$
$$x = 10,000.$$

Interpretation: Betty should deposit $10,000.

2. A couple wishes to move to a new subdivision where the lots cost $5,000 each and the houses cost $25 per square foot. The couple decide they can spend a total of $60,000 for the lot and house. What size house can they buy?

Erroneous Solution: If x denotes the number of square feet in a house and y denotes the total cost for the house and lot, the *mathe-*

matical model that expresses the relationship between x and y is given by

$$y = 25x.$$

Since the couple has $60,000 to spend, let y be 60,000 and *calculate* the corresponding value of x:

$$60,000 = 25x$$

$$\frac{60,000}{25} = x$$

$$x = 2,400.$$

Interpretation: The couple can buy a house with 2,400 square feet.

3.† Joan's yearly salary is $7,500. Next year she is scheduled for a 6% cost-of-living raise. If 36% of her salary is withheld each year for various deductions, such as income tax and social security, what will her take-home pay be next year?

Erroneous Solution: If x denotes Joan's present gross salary and y denotes her take-home pay for next year, the mathematical model that expresses the relationship between x and y is given by

$$y = x + 0.06x.$$

Because her present salary is $7,500, let x be 7,500 and *calculate* the corresponding value of y:

$$y = 7,500 + 0.06(7,500)$$
$$y = 7,500 + 450 = 7,950.$$

Interpretation: Joan's take-home pay for next year will be $7,950.

4. At a university on a 4.0 grading system, grade points for each semester hour of course work are awarded as follows:

Grade	Grade Points per Semester Hour
A	4
B	3
C	2
D	1
F	0

After 100 hours of course work, Ann has a grade point average of 1.96—that is, she has accumulated 196 grade points. Ann is taking a 3-hour summer course, hoping to raise her grade point average to 2.0. What grade must she make in the course in order to achieve this?

Erroneous Solution: Let y be Ann's grade point average at the end of the summer course and x the number of grade points per hour she earns in the summer course. The grade point average is the total number of grade points earned divided by the number of hours taken. Therefore, the *mathematical model* that expresses the relationship between x and y is

$$y = \frac{196 + 3x}{103}.$$

Let y be 2 and *calculate* the corresponding value of x:

$$2 = \frac{196 + 3x}{103}$$
$$206 = 196 + 3x$$
$$10 = 3x$$
$$x = 3 \text{ and } \tfrac{1}{3}.$$

Interpretation: If Ann earns a B in the summer course, she will have a grade point average of 2.0.

5.[†] A penny is dropped from the top of a building 400 feet tall. How long does it take the penny to reach the ground?

Erroneous Solution: Let d denote the distance the penny has fallen in time t. The *mathematical model* that expresses the relationship between d (measured in feet) and t (measured in seconds) is

$$d = 16t^2,$$

if the air resistance is ignored. We are interested in determining the time at which the distance is 400; so let d be 400 and *calculate* the corresponding value of t:

$$400 = 16t^2$$
$$25 = t^2$$
$$t = 6.$$

Interpretation: The penny will hit the ground in approximately six seconds. (Assume the given mathematical model is correct.)

6. A chemical company produces a saline solution (salt and water) that is normally 5% salt. Through error, a batch of eighty gallons was produced which contains only 3% salt. How much water should be evaporated from the solution to bring the salt content to 5%?

Erroneous Solution: Let x denote the amount of water to be evaporated off. The amount of remaining solution is then $80 - x$. The amount of salt will be the same before as after the evaporation, and the *mathematical model* that expresses this equality is

$$0.05(80 - x) = 0.03(80)$$

from which the value of x is *calculated*:

$$4 - 0.05x = 2.4$$

$$x = \frac{4 - 2.4}{0.05} = 32.$$

Interpretation: A 5% solution can be obtained by adding 32 pounds of salt.

4.3 RELATIONSHIP BETWEEN PURE AND APPLIED MATHEMATICS

In Chapter 1, it was pointed out that mathematics is generally divided into two classifications—pure and applied—depending upon whether the development of mathematical systems or the construction of appropriate mathematical models was the major concern. It was also pointed out that pure and applied mathematics cannot be separated into distinct categories with little or no interaction between the two.

The need for mathematical systems in applications of mathematics should be obvious from the previous sections. Each mathematical model considered was developed in the framework of a mathematical system, and the mathematical solution was obtained within that system. The reader should have been convinced that mathematical systems, the development of which is the primary objective of pure mathematics, are essential to applications.

On the other hand, the application of mathematics is often the motivating factor in the development of pure mathematics. A need for the theoretical apparatus to solve a particular problem indicates the need for the development or extension of a particular system. The development of graph theory as a result of Euler's desire to solve the Königsberg bridge problem is a perfect example. Thus, pure mathematics and applied mathematics advance together.

Another example of a mathematical system that was developed because of the need for mathematical models is *projective geometry*. Somewhat surprising, perhaps, is the fact that the problem from which projective geometry arose was the problem of perspective for the artist.

In Figure 4.4, a clock and checkerboard sit on the table. The face of the clock is no longer perfectly round, and the squares of the checkerboard are distorted so that they are no longer square. How can you recognize the objects when they are so distorted? The answer is that the picture is drawn in the same way as the physical objects would be seen by human eyes. The question that an artist painting such a picture must resolve is, "Which

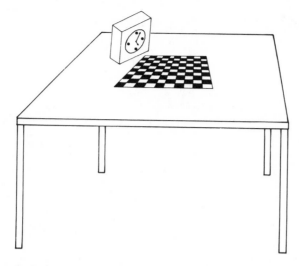

FIGURE 4.4

characteristics should be distorted and which should be retained in order to obtain a realistic picture?" By "a realistic picture" is meant a picture which depicts a scene in the same manner as the human eye sees it.

This question was first posed by the Renaissance artists, among them that Renaissance genius, Leonardo da Vinci (Fig. 4.5). The medieval painters before them rendered a more stylistic type of painting. with no great concern for the realism of the scenes depicted. The Renaissance artists, in the spirit of their times, wanted their works to depict reality. Because of this desire, they posed the question in the last paragraph. The artists of the time were also mathematicians, so they searched for a mathematical solution.

For a mathematical interpretation of the artists' problem, consider Figure 4.6. The line from point A to point O intersects the inclined plane at point A'. Similarly the lines from points B, C, D, and E to point O determine the points B', C', D', and E' in the inclined plane. Imagine that lines to point O are drawn from every point of the figure in the base plane. The points at which these lines intersect the inclined plane determine the figure

Leonardo da Vinci (1452–1519)

FIGURE 4.5

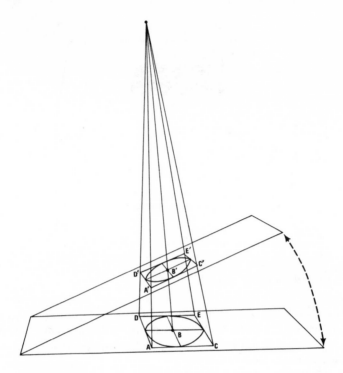

FIGURE 4.6

on the inclined plane. This figure is the *projection* in the inclined plane of the figure in the base plane.

Compare the figure with its projection in the inclined plane. Some characteristics are retained; such characteristics are called *invariants*. Because straight lines in the figure determine straight lines in the projection, straight lines are invariant. The intersections of lines and curves are also invariant.

Other characteristics are not retained. Compare the angle of intersection of straight lines, the area, and the parallel lines of the figure and its projection. Note also the shape of the projection of the circular portion of the figure.

If the point O were moved or if the inclination of the plane were changed, the projection of the figure would change also. Projective geometry is the study of those properties which remain invariant under all such projections.

How does this relate to the problem of the artist? Replace the point O by the human eye, the straight lines from the figure to the point O by light rays, and the inclined plane by a transparent sheet. The light rays determine an invisible projection on the transparent sheet. It is this projection that the artist attempts to paint. The question of which characteristics of the object should be preserved on the transparent sheet is answered by projective geometry.

It is not our purpose to study the full development of projective geometry. The point is that the system arose in the pursuit of a mathematical solution of a physical problem. It has been developed beyond the point of satisfying the need of the artists and has entered into applications in other areas such as relativity theory. Through such interaction, pure mathematics and applied mathematics exert a continuing influence on each other.

4.4 SUMMARY

The application of mathematics to solving physical problems involves the following four-step process.

1. A question is asked concerning a physical situation.
2. A mathematical model is constructed to describe the physical situation. The question in the first step now becomes a mathematical question.
3. The mathematical solution is found; that is, the mathematical question of Step 2 is answered.
4. The mathematical solution is interpreted back in the physical situation; that is, the question of Step 1 is answered.

If an appropriate mathematical model is not readily available, then a considerable amount of experimental work and trial and error is usually involved in constructing such a model.

The construction of a mathematical model and the determination of the mathematical solution are always done within the framework of some mathematical system. In this way, pure mathematics supplies the mathematical environment within which applied mathematics operates. On the other hand, applied mathematics influences the development of pure mathematics by indicating where mathematical theory is required for solving physical problems.

The Climatron: The Missouri Botanical (Shaw's) Garden, St. Louis, Missouri. One of the geodesic domes designed by their inventor R. Buckminster Fuller. The skeleton framework of *triangles* and *polygons* is so constructed that no interior supports are required.

chapter 5

analytic geometry— geometric models

Joan has a small company that makes clay pots. The process for making the pots requires that they be baked in a heated oven. A batch or lot of pots is placed in the oven for several hours; during this time the oven is kept at a constant temperature. Increasing fuel costs have made Joan very conscious of the expense involved in heating the oven.

Too much heat is wasted when a lot contains just a few pots. At the other extreme, a considerable amount of additional heat is required to maintain the oven at the correct temperature when the lot size is very large. Somewhere in between should be the ideal lot size at which the cost per pot is minimal. But what is this ideal size?

Joan carefully calculated the amount of gas used for lot sizes of 10, 17, and 25 pots. She then calculated the cost per pot and from her calculations made a chart of these costs (Fig. 5.1). The fuel cost was $1.35 per pot when the lot size was 10; the cost was 44¢ when the lot size was 17, and 60¢ when the lot size was 25. The higher costs appear higher in her chart.

To get an idea of what the cost might be for other lot sizes, she connected these costs with a curved line (Fig. 5.2). Based on the chart in Figure 5.2, she concluded that the minimal cost would be obtained if the lot size was about 20 or 21 pots.

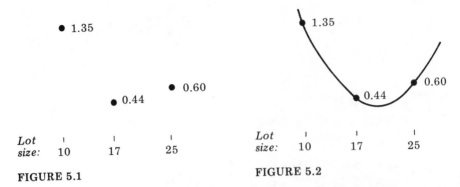

FIGURE 5.1   FIGURE 5.2

In Figure 5.2, Joan obtained a new type of mathematical description of her fuel costs—a geometric description. It is the purpose of this chapter to explore this idea of obtaining geometric descriptions or geometric models. In particular, we will seek geo-

metric representations of algebraic relationships. The setting within which such
representations are obtained is generally
referred to as analytic geometry.

Historically, the development of
analytic geometry is due to the French
philosopher and mathematician René
Descartes (Fig. 5.3). His discovery
turned out to be a happy marriage between geometry and algebra. As a consequence of analytic geometry, advances
in both geometry and algegra were made
that might never have occurred without
this union. The process by which this
union was accomplished is so simple that
one must wonder why it was not discov-
ered earlier.

René Descartes (1596–1650)

FIGURE 5.3

5.1 THE REAL-NUMBER LINE AND CARTESIAN PLANE

Analytic geometry begins with the construction of the real-number line, whereby the points on a line are placed in one-to-one correspondence with the real numbers.

On a horizontal straight line pick two points (Fig. 5.4). Label the point on the left 0 and the point on the right 1. The point labeled 0 is called the *origin*, and the distance between the points is the *unit of distance*.

FIGURE 5.4

Every point on the line to the right of the origin is labeled with the positive real number that is equal to the number of units that point is distant from the origin (Fig. 5.5).

FIGURE 5.5

Every point on the line to the left of the origin is labeled with the negative of the number that is equal to the number of units that point is distant from the origin (Fig. 5.6).

FIGURE 5.6

In this way, every point on the line is labeled with a real number, positive if the point is to the right of the origin and negative if the point is to the left of the origin. Conversely, every real number is the label of some point on the line.

The number that is associated with any point on the line is called the *coordinate* of that point. The number 3 is not the same as the point whose coordinate is 3, for example, but standard mathematical terminology does not distinguish between the two. "The point whose coordinate is 3" is shortened to "the point 3." In general, "the point whose coordinate is x" is replaced by the more simple phrase "the point x," for any real number x.

EXERCISE

1. Construct a real-number line, indicating the location of the following points: $\frac{1}{2}$, 2.3, 3.5, 5, π, −0.7, −2.5, −3, −$\sqrt{2}$, −5. (*Note:* π is approximately 3.1; $\sqrt{2}$ is approximately 1.4.)

Just as the points of a line can be put into a one-to-one association with the real numbers, the points of a plane can be put into a one-to-one association with pairs of real numbers. A plane is the surface of a sheet of paper or of a blackboard that extends indefinitely in all directions.

To establish this association, construct two real number lines at right angles to each other and intersecting at their origins (Fig. 5.7). The two number lines are called the *coordinate axes*, the horizontal axis being the *x-axis* and the vertical axis being the *y-axis*. The arrows indicate the positive directions.

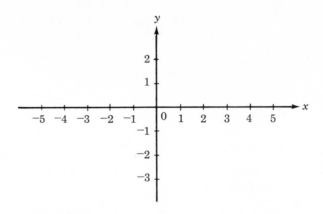

FIGURE 5.7

Every point in the plane is associated with a pair of numbers, (x,y), in the following way (Fig. 5.8). Through any given point, draw the line perpendicular to the x-axis; the coordinate of the point at which the line intersects the x-axis is the *x-coordinate* of the point. Through the same point, draw the line perpendicular to the y-axis; the coordinate of the point of intersection with the y-axis is the *y-coordinate* of the point. The pair of real numbers (x,y) determined in this manner are the *coordinates* of the point.

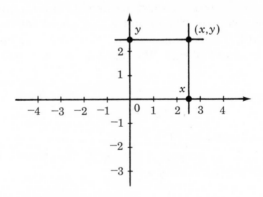

FIGURE 5.8

A plane with coordinates assigned to each point in this manner is called a *Cartesian plane* or a *coordinate system*. The location of some representative points and the coordinates of these points are shown in the coordinate system in Figure 5.9.

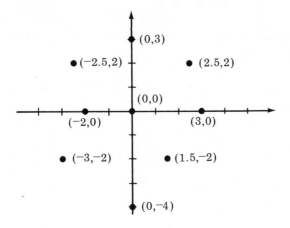

FIGURE 5.9

Conversely, if one starts with a pair of real numbers (x,y), the point with these coordinates can be determined by reversing the above process as follows. Through the point x on the x-axis draw the line perpendicular to the x-axis; through the point y on the y-axis draw the line perpendicular to the y-axis. The point of intersection of these two lines is the point having the coordinates (x,y).

EXAMPLE 1.

Locate the point with coordinates $(-2,3)$ in a coordinate system.

First draw the line perpendicular to the x-axis at the coordinate -2 (Fig. 5.10); then draw the line perpendicular to the y-axis at the coordinate 3. The point of intersection of these two lines is the desired point.

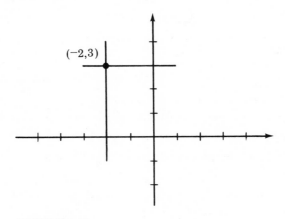

FIGURE 5.10

EXAMPLE 2.

Where are all of the points whose x-coordinate is 4 located in a co-ordinate system?

Because the value of the y-coordinate is immaterial, this would be all of the points on the line drawn through the point (4,0) and drawn perpendicular to the x-axis (Fig. 5.11).

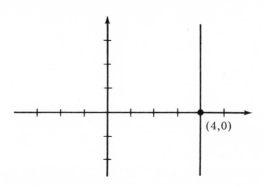

FIGURE 5.11

Just as with the real-number line, the point with coordinates (x,y) is distinct from the pair of real numbers (x,y). This distinction is usually ignored, and the phrase "the point with coordinates (x,y)" is usually shortened to "the point (x,y)."

EXERCISES

1.† Locate the following points in a coordinate system:

(1,1), (2,−3), (−3,2), (−1,−1.5), (0,2), (2,0), $(0,-\frac{1}{2})$, $(-\frac{1}{2},0)$.

2. Where are all of the points whose x-coordinate is −3 located in a coor-dinate system?

3.† Where are all of the points whose y-coordinate is 2 located in a coordi-nate system?

4. Where are all of the points whose x-coordinate is 0 located in a coordi-nate system?

5.† Where are all of the points whose y-coordinate is 0 located in a coordi-nate system?

6. Where are all of the points whose x-coordinate is equal to its y-coordi-nate located in a coordinate system?

5.2 GRAPHS OF EQUATIONS

The *graph* of an equation is the set of all points (x,y) whose coordinates satisfy the equation. For example, the point $(2,5)$ is in the graph of the equation $y = 2x + 1$ because, when x is given the value 2 and y is given the value 5 in the equation, equality holds: $5 = 2(2) + 1$. On the other hand, the point $(-2,0)$ is not in the graph of $y = 2x + 1$ because, when x and y are given the values -2 and 0, respectively, equality does not hold: $0 \neq 2(-2) + 1$.

To graph an equation means to locate all of the points (x,y) whose coordinates satisfy the equation. Generally, this means locating an infinite number of points in the plane. Consequently, it is impossible to locate each point individually. Instead, a few points are located, and these are then connected by a curve. The following examples illustrate this procedure.

EXAMPLE 1.

Graph the equation $y = x^2 + 1$.

We pick several values for x, say $x = 0, 1, 2, -1, -2$. For each of these values of x, we determine the corresponding value of y from the equation $y = x^2 + 1$. This pairing of values of x and y is represented in tabular form:

x	y
0	1
1	2
2	5
−1	2
−2	5.

Five points on the graph of $y = x^2 + 1$ are easily read from the table: $(0,1)$, $(1,2)$, $(2,5)$, $(-1,2)$, and $(-2,5)$; these points are then located in a coordinate system (Fig. 5.12). To complete the graph of the equation, these points are then connected by a curve (Fig. 5.13).

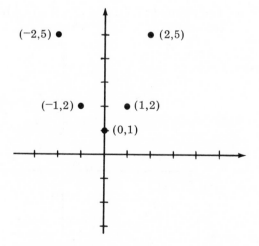

FIGURE 5.12

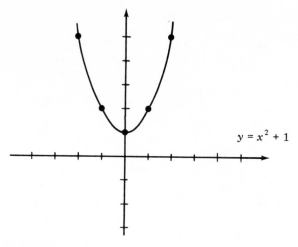

FIGURE 5.13

EXAMPLE 2.

Graph the equation $x + 2y = 3$.

We pick several values of x, say $x = -1, 0, 1, 2, 3$. The corresponding values of y are determined from the equation; in tabular form:

x	y
-1	2
0	$\frac{3}{2}$
1	1
2	$\frac{1}{2}$
3	$0.$

The points $(-1,2)$, $(0,\frac{3}{2})$, $(1,1)$, $(2,\frac{1}{2})$, and $(3,0)$ are then located and connected by a curve, which in this case is a straight line (Fig. 5.14).

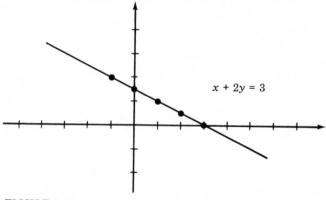

FIGURE 5.14

It is important to appreciate the relationship between an equation and its graph. The graph is the set of all those points, and only those points, whose coordinates satisfy the equation. By means of the graph, a geometric picture of the equation is obtained.

Obtaining the graph of an equation by the method considered in this section cannot be expected to be always completely accurate. After all, only a few points are used to determine the location of an infinite number. More sophisticated methods of graphing will be introduced in subsequent sections. Generally, this will be dependent upon an analysis of the *form* of the equation before any points are located.

EXERCISES

1.† Without constructing the graph, determine which of the following points are in the graph of the equation $y = 3x - 1$:

   (a) $(0,-1)$ (b) $(\frac{1}{3},0)$ (c) $(1,1)$
   (d) $(2,-4)$ (e) $(-1,-2)$ (f) $(-3,-10)$

2. Without constructing the graph, determine which of the following points are in the graph of the equation $y = 2x^2 + 2$:

   (a) $(1,2)$ (b) $(-1,4)$ (c) $(\frac{1}{2},0)$
   (d) $(-2,10)$ (e) $(2,-10)$ (f) $(0,-2)$

3. Choosing the x-coordinate to be from the set $\{-2, -1, 0, 1, 2\}$, determine five points in the graph of each of the following equations:

   (a)† $y = x^2$ (b) $y = 2x + 1$
   (c)† $y = -x^2 + 1$ (d) $2y = 2x - 1$
   (e)† $x - y = 3$ (f). $y - 1 = x^2$
   (g)† $y = x^3$

4.† Graph each of the equations in Exercise 3 by first locating the five points determined in Exercise 3, and then connecting the points with a curve.

5.3 EQUATION OF A STRAIGHT LINE

In the previous section, we started with an equation and then obtained the graph of the equation. The reverse approach is just as natural—given a graph, we can determine the equation of the graph. This latter approach will be used in this section to determine the equation when the graph is a straight line.

The basic concept in our approach will be the *slope* of a straight line. The slope of a straight line indicates the manner in which the line is slanted. Figure 5.15 shows several lines, all passing through the point (x_0, y_0). It is the direction in which each line is slanted that distinguishes it from all of the other lines.

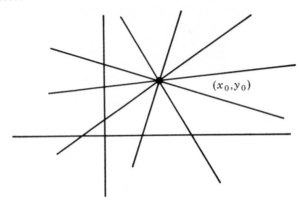

FIGURE 5.15

To define the slope of a straight line (except one parallel to the y-axis) mathematically, choose any two points, say (x_1, y_1) and (x_2, y_2) on the line. If we complete the right triangle determined by the points (x_1, y_1) and (x_2, y_2) in the manner indicated in Figure 5.16, the coordinates of the third vertex of the triangle will be (x_2, y_1).

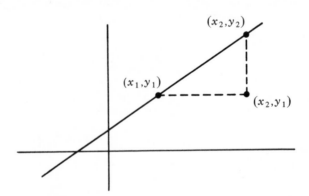

FIGURE 5.16

The length of the vertical side of the triangle is $y_2 - y_1$; the length of the horizontal side is $x_2 - x_1$. The *slope* of the given line is the quotient of these two lengths,

$$\frac{y_2 - y_1}{x_2 - x_1}.$$

Think of the point (x_2,y_2) as being obtained by moving from the point (x_1,y_1) along the line to (x_2,y_2). This quotient represents the ratio of the change in the y-coordinates to the change in x-coordinates as you move along the line from (x_1,y_1) to (x_2,y_2).

The slope concept is often used by engineers and architects to measure steepness. The pitch of a roof or the slope of a hill are measured in this way. In these contexts, the vertical change $(y_2 - y_1)$ is referred to as the *rise* and the horizontal change $(x_2 - x_1)$ is referred to as the *run*. The pitch of a roof, for instance, is the rise divided by the run.

The standard notation for the slope of a straight line is the letter m; that is,

$$m = \frac{y_2 - y_1}{x_2 - x_1}.$$

EXAMPLE 1.

Determine the slope of the straight line passing through the points $(-4,-1)$ and $(1,3)$.

Let $(x_2,y_2) = (1,3)$ and let $(x_1,y_1) = (-4,-1)$. Then,

$$m = \frac{y_2 - y_1}{x_2 - x_1} = \frac{3 - (-1)}{1 - (-4)} = \frac{4}{5}.$$

The chief feature of the slope definition that makes it so useful is the fact that the value of m does not change with the choice of the points (x_1,y_1) and (x_2,y_2). To see that this is the case, pick any two other points on a line ℓ, say (x_3,y_3) and (x_4,y_4) (Fig. 5.17). The angles of the triangle determined by the points (x_3,y_3) and (x_4,y_4) are the same as the angles of the triangle determined by (x_1,y_1) and (x_2,y_2). In geometry,

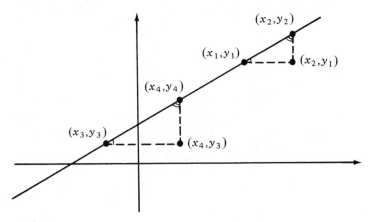

FIGURE 5.17

such pairs of triangles are called *similar triangles,* and a theorem of geometry states that the ratio of the lengths of their corresponding sides are equal. Hence,

$$\frac{y_4 - y_3}{x_4 - x_3} = \frac{y_2 - y_1}{x_2 - x_1}.$$

If a line is slanted upward to the right (as the one in Figure 5.17), both $y_2 - y_1$ and $x_2 - x_1$ are positive numbers. Their quotient is also a positive number. Consequently, such lines have a positive slope.

If a line is parallel to the x-axis, the y-coordinates of any two of its points will be the same; that is, $y_2 - y_1 = 0$. Such lines have a slope of 0.

If a line slants downward to the right (Fig. 5.18), $y_2 - y_1$ is negative, but $x_2 - x_1$ will be positive. The quotient $(y_2 - y_1)/(x_2 - x_1)$ will therefore be negative, and such lines have a negative slope.

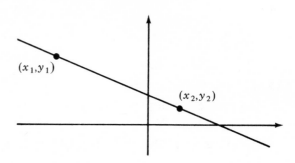

FIGURE 5.18

If a line is parallel to the y-axis, $x_2 - x_1 = 0$. Because division by zero has no meaning, such lines have no slope.

Some representative lines, all passing through the point (2,1), are given along with their slopes in Figure 5.19.

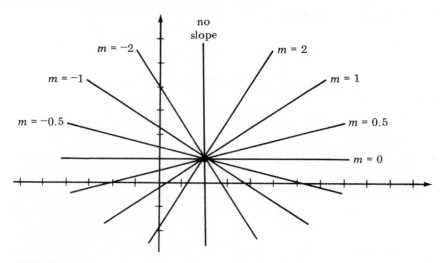

FIGURE 5.19

Note that

$$\frac{y_2 - y_1}{x_2 - x_1} = \frac{y_1 - y_2}{x_1 - x_2}.$$

As a consequence of this equality, it is immaterial which of two given points is chosen to be (x_1, y_1) and which is chosen to be (x_2, y_2) in calculating the slope of a line. The value of the resulting quotient will be the same.

EXAMPLE 2.

In a coordinate system, draw the line passing through the point $(-1,2)$ with slope $-\frac{3}{2}$.

Starting at the point $(-1,2)$ in Figure 5.20 we let x increase by 2 units. Because the slope is $-\frac{3}{2}$, the corresponding change in y is a decrease of 3 units. In this way, we have located another point on the line, which is then easily drawn.

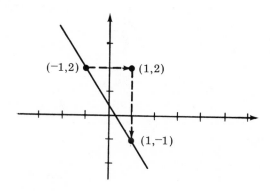

FIGURE 5.20

EXERCISES

1. Locate each of the following pairs of points in a coordinate system and draw the line which contains the two points. Then determine the slope (if there is one) of the line.

(a)† $(2,3)$ and $(-1,1)$
(b) $(2,-3)$ and $(1,-1)$
(c)† $(\frac{1}{2},0)$ and $(\frac{1}{4},1)$
(d) $(4,3)$ and $(-1,3)$
(e)† $(-3,4)$ and $(-3,5)$
(f) $(0.5,0.1)$ and $(0.6,2)$
(g)† $(17,14)$ and $(-18,14)$
(h) $(\frac{1}{3},-\frac{1}{3})$ and $(\frac{5}{6},\frac{7}{12})$

2. In a coordinate system, draw the line through the given point with the indicated slope.

(a)† Through (2,−3) with slope $\frac{3}{2}$.

(b) Through (4,1) with slope $-\frac{1}{3}$.

(c)† Through (0,0) with slope 1.

(d) Through (1,−1) with slope −2.

(e)† Through (−1,−2) with no slope.

Suppose now that we start with a given line and that we know the coordinates of one point (x_1, y_1) on the line as well as its slope m. How can we determine the equation whose graph is the given line?

The problem is to determine an equation such that the coordinates of each point on the line satisfy the equation. At the same time, the equation must be chosen so that the coordinates of no other point satisfy the equation.

Because any two points on the line determine the slope, each point (x, y) on the line must satisfy the equation

$$\frac{y - y_1}{x - x_1} = m.$$

At the same time, the coordinates of any point (x_0, y_0) not on the line do not satisfy this equation, as the points (x_1, y_1) and (x_0, y_0) determine a line with a different slope (Fig. 5.21).

The desired equation is therefore

$$\frac{y - y_1}{x - x_1} = m.$$

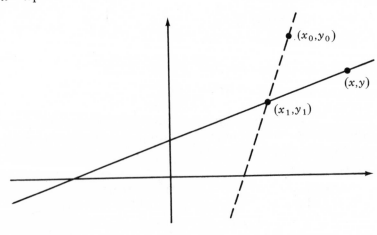

FIGURE 5.21

EXAMPLE 3.

Determine the equation of the line passing through the point (2,3) with slope 5.

We know $(x_1, y_1) = (2,3)$ and $m = 5$, so the equation is

$$\frac{y-3}{x-2} = 5.$$

To put the equation into a simpler form, multiply both sides by $(x - 2)$ to obtain

$$y - 3 = 5(x - 2).$$

Performing the indicated multiplication on the right side of this last equation gives

$$y - 3 = 5x - 10$$

or

$$y = 5x - 7.$$

_____

As in Example 3, the equation

$$\frac{y - y_1}{x - x_1} = m$$

can be written in a simpler form by multiplying each side of the equation by the quantity $x - x_1$ to obtain

$$\boxed{y - y_1 = m(x - x_1)},$$

which is usually referred to as the point-slope form of the equation of a straight line.

EXAMPLE 4.

Determine the equation of the line which passes through the two points (3,−1) and (4,3).

The slope of the line is

$$m = \frac{3 - (-1)}{4 - 3} = \frac{4}{1} = 4.$$

Which of the two points, (3,−1) or (4,3), is chosen to be the point (x_1, y_1) is immaterial, because both are points on the line. Using the point-slope form, with $(x_1, y_1) = (3,-1)$, the equation is

$$y + 1 = 4(x - 3)$$

or

$$y = 4x - 13.$$

EXAMPLE 5.

Determine the equation of the line which passes through the points $(5, -1)$ and $(3, -1)$.

The slope of the line is

$$m = \frac{-1 + 1}{5 - 3} = 0.$$

Choosing the point (x_1, y_1) to be $(5, -1)$, the equation is

$$y + 1 = 0$$

or

$$y = -1.$$

EXAMPLE 6.

Determine the equation of the line which passes through the points $(2,1)$ and $(2,-2)$.

$$x_1 - x_2 = 2 - 2 = 0.$$

Therefore, the line has no slope! The graph of the line in Figure 5.22 shows that an equation that is satisfied by the coordinates of all points on the line and by the coordinates of no other point is $x = 2$.

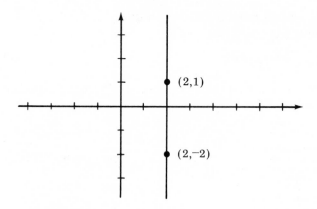

FIGURE 5.22

The point-slope form can be rewritten

$$y - y_1 = mx - mx_1$$

or

$$y = mx - mx_1 + y_1.$$

If b denotes the quantity $(-mx_1 + y_1)$, this last equation can be written

$$\boxed{y = mx + b}\,,$$

which is the slope-intercept form of the equation of a straight line. Note that if x is given the value 0, then $y = b$; that is, the line intersects the y-axis at the point $(0,b)$ (Fig. 5.23).

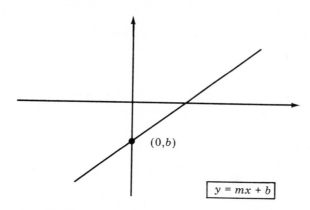

FIGURE 5.23

EXERCISES

1. Verify that the same equation is obtained in Example 4 of this section if the point (x_1, y_1) is taken to be $(4,3)$.

2. Determine the equation of the line
   (a)† through the point $(2,3)$ with slope 2.
   (b) through the point $(2,3)$ with slope -1.
   (c)† through the point $(\frac{1}{4},1)$ with slope -4.
   (d) through the point $(-4,3)$ with slope 0.
   (e)† through the points $(-3,3)$ and $(-3,5)$.
   (f) through the points $(1,4)$ and $(-1,8)$.
   (g)† through the points $(7,14)$ and $(-8,14)$.
   (h) through the points $(2.2,-1.1)$ and $(3.2,5.9)$.

3. Using the slope-intercept form, determine the equation of the line that
   (a)† has slope 3 and intersects the y-axis at the point $(0,-2)$.
   (b) has slope -2 and intersects the y-axis at the point $(0,0)$.
   (c)† passes through the point $(0,1)$ with slope -4.
   (d) passes through the point $(0,-4)$ with slope 1.

5.4 LINEAR EQUATIONS

A *linear equation* is any equation which can be written in the form
$Ax + By + C = 0$, where A, B, and C are real numbers, and A and B are
not both zero. (If A and B were both 0, the equation would become
$C = 0$ or $0 = 0$.)

Examples of linear equations are:

$$3x + 2y - 4 = 0$$

which can be written as $3x + 2y + (-4) = 0$, where $A = 3, B = 2$, and
$C = -4$;

$$-2x = 4y + 1$$

which can be written as $-2x - 4y - 1 = 0$ by subtracting $4y + 1$ from
each side; then $A = -2, B = -4$, and $C = -1$;

$$0.5x + \frac{y}{2} - \frac{1}{4} = 0$$

where $A = 0.5, B = \frac{1}{2}, C = -\frac{1}{4}$; and

$$3y = -2.$$

What are A, B, and C in this last equation?

The important thing to note about the form of a linear equation is
that the highest power in which x and y appear is the first power; there
can be no x^2, y^2, or any higher power. Nor are there any products involving both x and y, such as xy, or xy^2.

The equation of every straight line can be written in the form
$y = mx + b$; or, if there is no slope, in the form $x = c$. It therefore follows
that the equation of every straight line is a linear equation. Conversely,
the graph of every linear equation is a straight line—hence the term *linear*
equation. Note that the equations of Exercises 3(b), 3(d), and 3(e) of
Section 5.2 are linear, and that their graphs are straight lines.

To verify the statement that the graph of every linear equation is a
straight line, suppose first that B is not zero in $Ax + By + C = 0$. Then
division by B is possible, and the equation can be written

$$y = -\frac{A}{B}x - \frac{C}{B}.$$

The graph of this equation is the line passing through $(0,-C/B)$ with slope
$-A/B$.

If B is zero, then A cannot be zero, and the general equation is
$Ax + C = 0$, which can be rewritten

$$x = -\frac{C}{A},$$

which has as its graph the line parallel to the y-axis and passing through the point $(-C/A,0)$. Consequently, the graph of a linear equation is always a straight line, whether or not B is zero.

The knowledge of this fact makes it particularly easy to graph any linear equation. The location of two points on any line determine completely the location of the line. Therefore, it is sufficient to locate only two points, and then to draw the line through them, to obtain the graph.

EXAMPLE 1.

Graph the equation $3x - y = 4$.

Choose any two values of x, say $x = 0$ and $x = 1$. If x is set equal to zero in the equation, $y = -4$; if x is set equal to 1, $y = -1$. Therefore, $(0,-4)$ and $(1,-1)$ are two points in the graph (Fig. 5.24).

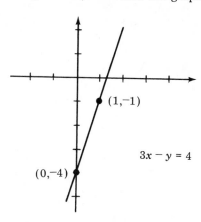

FIGURE 5.24

EXAMPLE 2.

Determine the slope of the line whose equation is $2x + 3y = 4$.

Subtracting $2x$ from each side of the given equation, we have

$$3y = -2x + 4.$$

Dividing both sides by 3 gives

$$y = -\tfrac{2}{3}x + \tfrac{4}{3},$$

which is in the form $y = mx + b$. The slope is therefore $-\tfrac{2}{3}$.

EXERCISES

1. Graph each of the following equations.
   (a)† $x - 3y = -2$ (b) $2x + y = 0$
   (c)† $y = 2x + 3$ (d) $-3x - y + 1 = 0$
   (e)† $x + y = 0$ (f) $x = y$
   (g)† $x = 1.5$ (h) $y = -1.5$

2.† Determine the slope of each of the lines whose equations are given in Exercise 1.

5.5   LINEAR EQUATIONS AS MATHEMATICAL MODELS

When a physical problem involves a relationship between two quantities and the mathematical model that expresses this relationship is an equation, the graph of the equation gives a second mathematical model—a geometric model. When the equation is a linear equation, the geometric model is, of course, a straight line.

By way of illustration, consider Frank, the corner grocer. He has found that his operating expenses each month amount to $700 plus 13% of his sales. Do his operating expenses increase rapidly, moderately, or slowly with sales?

Frank's operating expenses (y) in terms of sales (x) are $y = 700 + 0.13x$. This linear equation is the mathematical model of his operating expenses. Its graph gives a geometric model (Fig. 5.25). The relatively small slope of the line indicates that operating expenses increase slowly with sales.

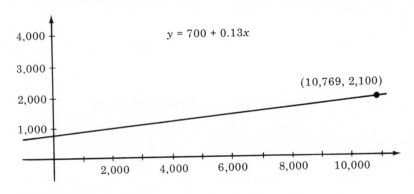

FIGURE 5.25

If Frank's operating expenses for a particular month amounted to $2,100.00, then his total sales for that month can be determined by solving the equation

$$2,100 = 700 + 0.13x$$

to obtain $x = 10,769$ (approximately). The corresponding point is located on the graph in Figure 5.25.

The unit distance in the coordinate system of Figure 5.25 was chosen to be very small. This was done for convenience, because the numbers under consideration were relatively large. Note that the points of the graph with negative x-coordinates are meaningless from the point of view of the expense problem; after all, how can one have negative sales!

Several of the mathematical models that you were asked to construct in Section 4.1 were linear equations. Figure 5.26 and the exercises at the end of this section are based on the equations constructed at that time.

Consider Herman, the builder of Exercise 7, Section 4.1. The total cost (y) of the trunk in terms of the number of square feet of plywood used (x) is $y = 3 + (\frac{1}{2})x$. Does the cost of the trunk increase rapidly, moderately, or slowly with the increase in the amount of plywood used? If he decides to use 40 square feet of plywood, what is the cost of the trunk?

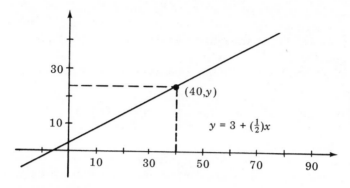

FIGURE 5.26

Because the slope of the line (the geometric model) is moderate, the increase in cost is moderate as the amount of plywood increases.

The cost of the trunk, if 40 square feet of plywood is used, can be determined by finding the y-coordinate of the corresponding point on the graph in Figure 5.26. However, this type of question can be answered more accurately by solving the corresponding equation—in this case $y = 3 + (\frac{1}{2})x$ with $x = 40$.

Again, the points in the graph with negative x-coordinates have no meaning in this situation.

EXERCISES

1.† (a) Graph the equation of Exercise 1, Section 4.1, that relates the number of books stolen to the number of books placed on the shelves.

  (b) If 500 books are shelved during a particular month, locate the corresponding point in the graph.

  (c) Judging from the graph in part (a), would you say that the number of thefts increases slowly, moderately, or quickly with the number of books placed on the shelves?

  (d) What points in the graph are relevant to this particular situation? (*Hint:* Consider realistic values of x.)

2. (a) Graph the equation which relates the number of defective bulbs to the total number of bulbs manufactured by the electrical supplies firm of Exercise 3, Section 4.1.

(b) If 10,000 light bulbs are manufactured during a particular week, locate the corresponding point in the graph.

(c) What points in the graph are relevant to this particular situation?

3.† (a) Graph the equation that expresses the number of reminder notices required in terms of the number of bills mailed by the credit manager of Exercise 5, Section 4.1.

(b) During March, the credit manager had to send out 250 reminder notices. Locate the corresponding point in the graph of the equation.

(c) What points in the graph are relevant to this particular situation?

4. (a) Graph the equation of Exercise 6, Section 4.1, that expresses the number of meters in any number of yards.

(b) A football field is 100 yards long. Locate the corresponding point in the graph.

(c) Judging from the graph of part (a), would you say that the number of meters increases slowly, moderately, or quickly with the number of yards?

5.† (a) Graph the equation that relates the salary of the basketball player of Exercise 9, Section 4.1, to the number of games won.

(b) If his salary was $51,000 for the year, locate the corresponding point in the graph.

6. (a) Graph the equation that relates Tom's salary to the number of television sets that he sells as indicated in Exercise 10, Section 4.1.

(b) During April, Tom sold 17 television sets. Locate the corresponding point in the graph.

7.† (a) Graph the equation that expresses the price per person in terms of the number of empty seats in the chartered flight of Exercise 13, Section 4.1. (To maintain consistent notation, let the *y*-axis be the *z*-axis.)

(b) If there are 83 people on the flight, locate the corresponding point in the graph.

(c) What points in the graph are relevant to this particular situation?

8. (a) Graph the equation that expresses the profit per tree in terms of the number of additional trees planted by the farmer of Exercise 14, Section 4.1.

(b) If the farmer plants 30 additional trees, locate the corresponding point in the graph.

9.† (a) Graph the equation that expresses the rent per apartment in terms of the number of empty apartments in the complex owned by Paul of Exercise 16, Section 4.1.

(b) If Paul has four empty apartments, locate the corresponding point in the graph.

(c) Which points in the graph are relevant to this particular situation?

5.6 SYSTEMS OF LINEAR EQUATIONS–APPLICATIONS

Susan owns and manages a large record store. At the end of each month, she replenishes her stock of records by ordering as many records as she sold during the month. During May, she sold 1,452 records and will therefore order that number of records.

Many of Susan's customers are classical music devotees. To provide a good selection for these customers, the number of classical records that she orders is in proportion to the number of classical records sold. She estimates that, during May, the number of classical records sold was twice the number of nonclassical. How many of the 1,452 records should be classical and how many nonclassical?

To construct a mathematical model describing Susan's record problem, let x denote the number of classical records and y the number of nonclassical records to be bought. Because the total number to be replaced is 1,452, x and y must satisfy the relation

$x + y = 1,452.$

In addition, because the number of classical records must be twice the number of nonclassical records, x and y must also satisfy the equation

$x = 2y.$

These two equations together form the mathematical model for solving Susan's problem. Their graphs are shown in Figure 5.27. Because

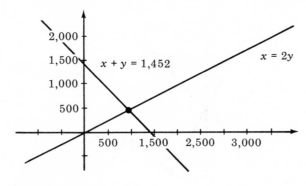

FIGURE 5.27

the numbers of records to be bought, x and y, must be the coordinates of a point on both of these graphs, they must be the coordinates of the point of intersection of these two lines.

The coordinates of the point of intersection can be approximated from the graph, or they can be obtained exactly by solving the pair of linear equations

$$x + y = 1,452$$
$$x = 2y.$$

To obtain this solution, substitute into the first equation the value of x obtained from the second. This gives

$$2y + y = 1,452$$

or

$$3y = 1,452,$$

which can easily be solved to obtain $y = 484$.

With this value for y, the proper value for x can easily be determined from the equation $x = 2y$ to be 968. It follows that Susan should buy 968 classical records and 484 nonclassical records.

Any such pair of equations is called a *system* of equations. A *solution* of such a system is any pair of values for x and y which satisfy both equations. Geometrically, the solution(s) of a system is (are) the points at which the graphs of the equations intersect.

EXAMPLE 1.

Solve the system of equations

$$2x + y = 1$$
$$x - y = -1.$$

The graphs of the two equations intersect in one point, the coordinates of which give the solution of the system (Fig. 5.28). To solve

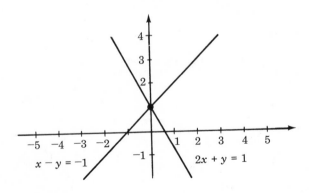

FIGURE 5.28

the system algebraically, rewrite the first equation to obtain y in terms of x: $y = 1 - 2x$. Substitute this expression for y into the second equation: $x - (1 - 2x) = -1$. This last equation can be rewritten $3x - 1 = -1$, from which it follows that $x = 0$.

If x is given the value zero in the first equation, it follows that $y = 1$. The solution is therefore $x = 0$ and $y = 1$.

The procedure by which the solution was obtained in this last example can be followed in solving any pair of linear equations:

1. Use one of the equations to express x in terms of y, or y in terms of x. (From the first equation of Example 1, y is expressed in terms of x: $y = 1 - 2x$.)
2. Substitute into the remaining equation the expression so obtained. (In Example 1, y is replaced in the second equation by the expression $1 - 2x$.)
3. Step 2 results in a linear equation which contains only x or only y. This equation can easily be solved to determine the value of the remaining unknown, x or y, in the solution. (In Example 1, the resulting equation, $x - (1 - 2x) = -1$, is solved to obtain $x = 0$.)
4. In either of the original equations, x or y (whichever is appropriate) is given its value determined in Step 3. There results a linear equation which can easily be solved for the remainder of the solution. (In Example 1, x is given the value zero in the first equation to obtain $y = 1$.)

In Example 1 and in the illustration of Susan's record shop, the graphs of the two equations intersected at one point, and the coordinates of that point were the solution of the system. Figure 5.29 gives two other possible combinations of the graphs of two linear equations. In the coordinate system on the left of Figure 5.29, the two lines are parallel and con-

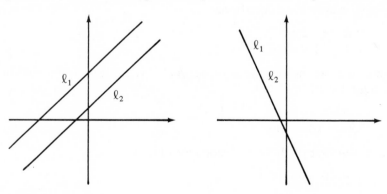

FIGURE 5.29

sequently never intersect. The equations of these two lines would form a system which has no solution.

In the coordinate system on the right of Figure 5.29, the two lines coincide. The coordinates of any point on the line form a solution of the system made up of the equations of the two lines. In this case, there are infinitely many solutions.

The four-step procedure outlined after Example 1 for solving a system of linear equations will reveal whether a system has one, none, or infinitely many solutions. In case of no solutions, the third step of the procedure results in an equation which can never be valid. In case of infinitely many solutions, the third step results in an equation which is always valid. The next two examples illustrate these two cases.

EXAMPLE 2.

Solve the system of equations

$$x - 3y = 2$$
$$9y - 3x = 4.$$

From the first equation, we have $x = 2 + 3y$. Replacing x by $2 + 3y$ in the second equation yields

$$9y - 3(2 + 3y) = 4$$

or

$$9y - 6 - 9y = 4.$$

This last equation implies

$$-6 = 4,$$

which can never hold for any value of y; hence, there is no solution.

EXAMPLE 3.

Solve the system of equations

$$0.2x - 0.3y = 1$$
$$1.2x - 1.8y = 6.$$

First, to clear the equations of decimals, we multiply each equation through by 10 to obtain

$$2x - 3y = 10$$
$$12x - 18y = 60.$$

From the first of these equations we obtain

$$x = \frac{10 + 3y}{2} \tag{5.1}$$

Replacing x by $(10 + 3y)/2$ in the second equation gives

$$12\left(\frac{10 + 3y}{2}\right) - 18y = 60,$$

$$6(10 + 3y) - 18y = 60,$$

or

$$60 + 18y - 18y = 60.$$

This last equation implies that $0 = 0$, which holds for all values of y. We can therefore give y any value and the corresponding value of x can be determined by Equation (5.1). For example, if y is given the value zero, then

$$x = \frac{10 + 3(0)}{2} = 5.$$

One solution is therefore $x = 5$ and $y = 0$.

If y is given the value 4,

$$x = \frac{10 + 3(4)}{2} = 11.$$

Hence, another solution is $x = 11$ and $y = 4$.

If the first equation had been solved for y, instead of x, to obtain

$$y = \frac{2x - 10}{3} \tag{5.2}$$

in place of Equation (5.1), then x can be assigned arbitrary values and the corresponding value of y will be determined by Equation (5.2). Suppose, for example, x is given the value 11, then

$$y = \frac{2(11) - 10}{3} = 4$$

and one solution is $x = 11$ and $y = 4$.

Either Equation (5.1) or Equation (5.2) can be stated as the solution of the system of equations.

EXAMPLE 4.

Solve the system of equations

$$\frac{x}{4} + \frac{2y}{5} = 4$$

$$\frac{y}{2} - \frac{x}{3} = 5.$$

To clear the equations of fractions, multiply each equation by the least common denominator of the fractions involved. For the first equation, the least common denominator is 20; multiplying each side by 20 gives

$$\left(20 \times \frac{x}{4}\right) + \left(20 \times \frac{2y}{5}\right) = 20 \times 4$$

or

$$5x + 8y = 80.$$

Similarly, multiply each side of the second equation by 6 to obtain

$$3y - 2x = 30.$$

The resulting pair of equations

$$5x + 8y = 80$$
$$3y - 2x = 30$$

can then be solved by the four-step procedure to obtain the solution $x = 0$ and $y = 10$.

EXERCISES

1. Solve the following systems of equations. (Graph each pair of equations, if you find that to do so is helpful.)

   (a)† $x + y = 1$
   $\quad\quad 3x - y = 7$

   (b) $\dfrac{x}{2} + \dfrac{y}{3} = 2$

   $\quad\quad \dfrac{x}{4} - \dfrac{y}{6} = 0$

   (c)† $0.3x + 0.9y = 0.3$
   $\quad\quad -1.2x - 3.6y = 0$

   (d) $\dfrac{y}{2} - \dfrac{x}{2} = 1$

   $\quad\quad x - y = 1$

   (e)† $y - x = 4$
   $\quad\quad 2y - 2x = 8$

   (f) $\quad 2x + 2y = 0$
   $\quad\quad 0.5x + 0.5y = 0$

2.† A certain manufacturing company produces two items, A and B. The manufacturing process emits 1.5 cubic feet of carbon monoxide and 3 cubic feet of sulfur dioxide per unit of item A, and 2 cubic feet of both carbon monixide and sulfur dioxide per unit of item B. Government pollution standards allow the company to emit a maximum of 4,775 cubic feet of carbon monoxide and 5,500 cubic feet of sulfur dioxide per week.

   (a) Let x denote the number of units of item A produced per week and y the number of units of item B. Write an equation to express the combined number of units of item A and of item B that will cause a total emission of 4,775 cubic feet of carbon monoxide.

(b) Using the same notation as in part (a), write an equation to express the combined number of units of item A and of item B that will cause a total emission of 5,500 cubic feet of sulfur dioxide.

(c) From the equations determined in parts (a) and (b), determine the number of units of items A and B which can be produced without exceeding government standards.

3. The cost of manufacturing stereo equipment of a particular type is given by

$$C = \$12{,}000 + \$120x$$

where x denotes the number of units produced. These units are sold for \$200 each; the revenue from x units is therefore

$$R = \$200x.$$

(a) In a coordinate system, graph the two equations

$$y = 12{,}000 + 120x$$

and

$$y = 200x$$

to represent cost and revenue.

(b) The point of intersection of the two graphs in part (a) is called the *break-even point*, as it represents the point at which revenue equals cost. Solve the system of equations to determine the coordinates of the break-even point.

(c) How many units must the company sell to break even?

(d) What must be the revenue received for the company to break even?

4.† (a) Of the 168 hours in a week, Marilyn decides to devote 60 hours to her classes and to her part-time job. Denote the number of hours that Marilyn spends per week on her classes by x, and denote the number of hours per week on her job by y. Write an equation to express x and y in terms of the number of hours available for the two activities.

(b) Marilyn believes that she should spend twice as much time on her classes as she does on her job. Write an equation in x and y to express this relationship.

(c) Solve the system of equations arrived at in parts (a) and (b) to determine how many hours should be spent on each activity.

(d) Marilyn is carrying a course load of 12 hours. How many hours does she have for studying and homework?

5. Bob owns a bicycle shop and wants to order 50 bicycles for the spring rush. Three-speed bicycles cost him $60 each, and ten-speeds cost $70 each. He has $3,200 with which to buy the bicycles.

   (a) Let x denote the number of three-speed bicycles to be bought, and let y represent the number of ten-speed bicycles. Write an equation in x and y to express the fact that a total of 50 bicycles are to be bought.

   (b) Write an equation in x and y to express the fact that $3,200 are to be spent on the bicycles.

   (c) Solve the system of equations obtained in parts (a) and (b) to determine how many of each type of bicycle Bob should buy.

*6. In solving a system of three equations involving three unknown numbers, one can first reduce the problem to a pair of equations involving two unknown numbers. Then the procedure discussed in this section is used to solve the pair of equations in the reduced system. This process is outlined in part (a) below.

   (a)† Solve the following system for the value of x, y, and z by following the procedure given below.

$$x + y + z = 7$$
$$3x + 2y - z = 3$$
$$3x + y + 2z = 13$$

   (i)   Add the first two equations to obtain an equation involving only x and y.

   (ii)   Multiply each side of the first equation by -2. Add the resulting equation to the third equation to obtain another equation in x and y.

   (iii) The equations obtained in Steps (i) and (ii) make up a system of equations in x and y. Solve this sytem by the method explained in this section.

   (iv)   Substitute the value of x and y obtained in Step (iii) into any one of the original equations. The result will be an equation in z alone. Solve this equation for the value of z.

   (b) Solve the system:

$$x + y + z = 2$$
$$2x - y + z = 5$$
$$-x + y - 3z = -8.$$

5.7 DISTANCE IN A COORDINATE SYSTEM

A concept basic to the remainder of our work in coordinate systems is that of the *distance* between two points. This distance, as we shall see, is easy to calculate in terms of the coordinates of the points.

Let the points be P_1, with coordinates (x_1, y_1), and P_2, with coordinates (x_2, y_2). Locate the points in a coordinate system and complete the right triangle determined by the two points (Fig. 5.30). The distance between points P_1 and P_2 is equal to the length of the line segment joining

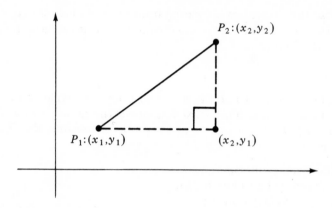

FIGURE 5.30

them. The length of this line segment can be determined by the Pythagorean theorem from geometry, which states that the square of the length of the hypotenuse, which is the side opposite the right angle, is equal to the sum of the squares of the lengths of the other two sides.

The lengths of the other two sides of the triangle in the graph can be determined by the coordinates of their endpoints. The horizontal side has length $x_2 - x_1$; its length squared is therefore $(x_2 - x_1)^2$. The vertical side has length $y_2 - y_1$, which when squared becomes $(y_2 - y_1)^2$.

According to the Pythagorean theorem, the square of the distance between P_1 and P_2 is given by

$$(x_2 - x_1)^2 + (y_2 - y_1)^2.$$

The square root of this last quantity gives the desired distance; that is, if we denote the distance between P_1 and P_2 by the symbol $\overline{P_1P_2}$, then

$$\boxed{\overline{P_1P_2} = \sqrt{(x_2 - x_1)^2 + (y_2 - y_1)^2}.}$$

EXAMPLE 1.

Determine the distance between the points $(2,5)$ and $(-1,1)$.

If P_1 is the point $(2,5)$ and if P_2 is the point $(-1,1)$ then

$$\overline{P_1P_2} = \sqrt{(-1-2)^2 + (1-5)^2}$$
$$= \sqrt{(-3)^2 + (-4)^2}$$
$$= \sqrt{9+16}$$
$$= \sqrt{25}$$
$$= 5.$$

The distance between the two points is therefore 5 units, where a *unit* is the distance from the origin to the point $(1,0)$.

Because the quantities appearing under the radical sign are squared, note that it makes no difference which point is taken to be P_1 and which is taken to be P_2.

EXAMPLE 2.

A *circle* is defined to be the set of all points which are equidistant from a fixed point. The equal distance is the *radius* of the circle and the fixed point is the *center* of the circle. Determine the equation of a circle.

Let the fixed point be (h,k) and the radius be r (Fig. 5.31). All the points (x,y) on the circle (and only those points) are at the distance r from (h,k); therefore, the desired equation is

$$r = \sqrt{(x-h)^2 + (y-k)^2}$$

or, squaring both sides,

$$r^2 = (x-h)^2 + (y-k)^2.$$

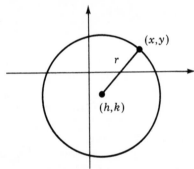

FIGURE 5.31

EXAMPLE 3.

Graph the equation
$4 = (x-2)^2 + (y+1)^2$.

In view of the equation derived in Example 2, the given equation can be rewritten in the form $2^2 = (x-2)^2 + [y-(-1)]^2$. Its graph is therefore the circle of radius 2 with center at $(2,-1)$, as shown in Figure 5.32.

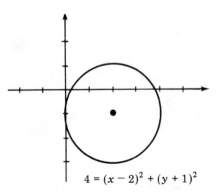

$4 = (x-2)^2 + (y+1)^2$

FIGURE 5.32

EXERCISES

1. Determine the distance between the following pairs of points:
   (a)† (2,3) and (8,11)
   (b) (−1,1) and (2,−3)
   (c)† (0,0) and (9,12)
   (d) (1,1) and (2,2)
   (e)† (−2,−2) and (0,−4)
   (f) (6,$\sqrt{3}$) and (3,0).

2. Graph the following equations:
   (a)† $9 = (x − 2)^2 + (y − 3)^2$
   (b) $25 = (x − 1)^2 + (y + 1)^2$
   (c)† $16 = (x + 2)^2 + (y + 2)^2$
   (d) $8 = (x − 2.5)^2 + (y + 4.2)^2$

5.8 PARABOLAS

Figure 5.33 depicts a cross section of an automobile headlamp. The broken lines depict light rays emanating from the light bulb in the center. Note that the light rays leave the headlamp in straight lines and therefore form

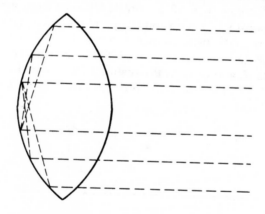

FIGURE 5.33

a concentrated beam, resulting in a more satisfactory light than if the rays left the headlamp in all directions. The cause for this concentrated beam, in conjunction with the behavior of light rays, is the shape of the reflector part of the headlamp. The cross section of the reflector is in the shape of a *parabola*.

Geometrically, a *parabola* is defined to be the set of all points that are equidistant from a fixed point and a fixed line. The fixed point is called the *focus* of the parabola and the fixed line the *directrix*. In Figure 5.34, suppose that the focus is (3,1) and the directrix is $y = −1$. Then the point

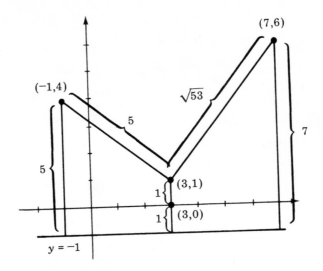

FIGURE 5.34

(3,0) belongs to the parabola, because its distance from the point (3,1) is 1 and its distance from the line $y = -1$ is also 1. The point $(-1,4)$ also belongs, because its distance from (3,1) and its distance from $y = -1$ are both 5. However, point (7,6) is not on the parabola; its distance from (3,1) is not the same as its distance from $y = -1$.

In general, if F denotes the focus and if D is the directrix, then point P is on the parabola if, and only if, the distance $\overline{PF}$ is equal to the distance from P to D (Fig. 5.35). Denote the coordinates of F by $(h, k+p)$ and the

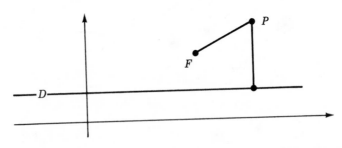

F IGURE 5.35

equation of D by $y = k - p$ (for reasons which we shall indicate shortly). Then, by the distance formula of Section 5.7, these two distances are equal if the coordinates (x,y) of P satisfy the equation

$$4p(y - k) = (x - h)^2 .$$

Its graph will take the form shown in Figure 5.36. The point (h,k), which is at a distance of p units from both the focus and the directrix, is called

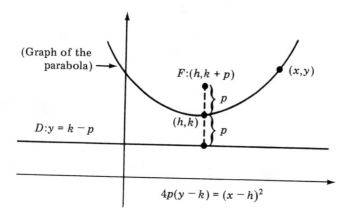

(Graph of the parabola) →

$F:(h,k + p)$

(x,y)

p

$D:y = k - p$

(h,k)

p

$4p(y - k) = (x - h)^2$

FIGURE 5.36

the *vertex* of the parabola. From the point of view of applications later in this chapter, the vertex is the most important point on the parabola.

Note that the coordinates of the vertex (h,k), as well as the distance p from the vertex to the focus and directrix, are given explicitly in the equation

$$4\underline{\underline{p}}(y - \underline{\underline{k}}) = (x - \underline{\underline{h}})^2$$

of the parabola. The seemingly odd choice of coordinates for the focus and of the equation for the directrix were made above to obtain precisely this result! The equation itself can be derived using the distance formula of the previous section.

From the point of view of graphing a parabola, it is important to realize that the location of the vertex, focus, and directrix can be determined once the values of h, k, and p are known.

EXAMPLE 1.

Locate the vertex, focus, and directrix and graph the parabola $8(y - 1) = (x - 3)^2$.

Use the general equation of a parabola previously derived, $4p(y - k) = (x - h)^2$. For this particular parabola, $p = 2$, $h = 3$, and $k = 1$. The vertex is therefore at $(3,1)$,

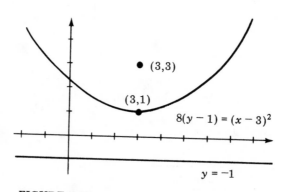

$(3,3)$

$(3,1)$

$8(y - 1) = (x - 3)^2$

$y = -1$

FIGURE 5.37

the focus is at $(h, k + p) = (3,3)$, and the equation of the directrix is $y = k - p = -1$. The graph is shown in Figure 5.37. Recall that the graph is the set of all points whose coordinates (x,y) satisfy the given equation.

EXAMPLE 2.

Determine the equation of the parabola with vertex at $(1,-1)$ and directrix $y = -4$.

Locate the given information in a coordinate system as indicated. The equation of the parabola will be in the form $4p(y - k) = (x - h)^2$. We have only to determine the values of h, k, and p. Because the vertex is at $(1,-1)$, $h = 1$ and $k = -1$. Because the distance from the vertex to the directrix is p, it can be seen in Figure 5.38 that $p = 3$. The desired equation is then

$$4(3)[y - (-1)] = (x - 1)^2$$

or

$$12(y + 1) = (x - 1)^2.$$

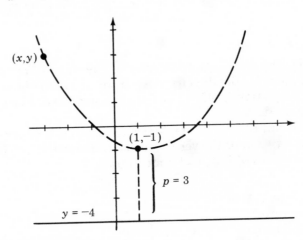

FIGURE 5.38

The equation $4p(y - k) = (x - h)^2$ is only for parabolas which open upward. There is a slight change for parabolas which open downward. If in the beginning the directrix had been chosen to be above the focus (Fig. 5.39) with equation $y = k + p$, the resulting equation would have been

$$-4p(y - k) = (x - h)^2 .$$

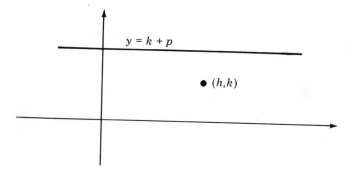

FIGURE 5.39

Its graph, which now opens downward, is given in Figure 5.40. The only difference in the form of the two equations is that the coefficient of $(y - k)$ is now a negative quantity; in the first case, it was positive.

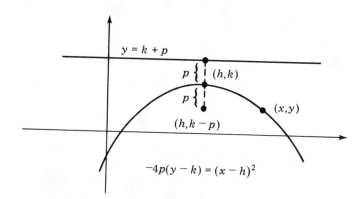

FIGURE 5.40

EXAMPLE 3.

Locate the vertex, focus, and directrix and graph the parabola $-8(y - 1) = (x - 3)^2$. (Compare with Example 1.)

In this case again, $p = 2$, $h = 3$, and $k = 1$, but the negative coefficient of $(y - 1)$ indicates that the parabola opens downward. The vertex is at $(3,1)$, the focus at $(3, k - p) = (3, -1)$ and the equation of the directrix is $y = k + p = 3$. The graph is given in Figure 5.41.

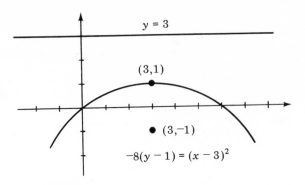

FIGURE 5.41

EXAMPLE 4.

Determine the equation of the parabola with vertex at $(1,2)$ and focus at $(1,-1)$.

The graph of the given information indicates that the focus is below the vertex (Fig. 5.42). Therefore, the equation is in the form $-4p(y-k) = (x-h)^2$.

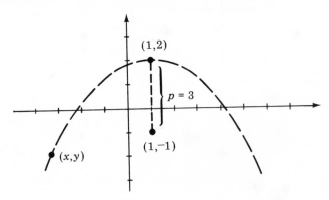

FIGURE 5.42

Because the vertex is $(1,2)$, it follows that $h = 1$ and $k = 2$. From the graph, we see that $p = 3$. The desired equation is therefore

$$-4(3)(y-2) = (x-1)^2$$

or

$$-12(y-2) = (x-1)^2.$$

There are an infinite number of parabolas opening upward or downward with the same vertex. They are distinguished one from the other by the value of p, which determines how widely each parabola opens. It is helpful, when graphing any particular parabola, to locate one point to the left and one to the right of the vertex; these points will indicate how widely the parabola opens.

EXAMPLE 5.

Graph the equation $6(y - 4) = (x + 3)^2$.

Because the coefficient of $(y - 4)$ is positive, the graph will be a parabola opening upward. The vertex will be at $(-3,4)$. Because $4p = 6$, $p = \frac{6}{4}$ or 1.5; the focus will be at $(-3,5.5)$. (See Fig. 5.43.)

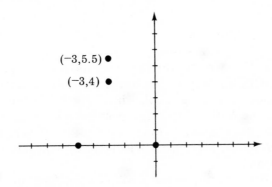

$(-3,5.5)$ ●
$(-3,4)$ ●

FIGURE 5.43

To determine how widely the parabola opens, calculate the value of y when $x = 0$ and when $x = -5$. For $x = 0$,

$$6(y - 4) = (0 + 3)^2,$$
$$6y - 24 = 9,$$
$$6y = 33,$$
$$y = \frac{33}{6} = 5.5.$$

For $x = -5$,

$$6(y - 4) = (-5 + 3)^2,$$
$$6y - 24 = 4,$$
$$6y = 28,$$
$$y = \frac{28}{6} = 4\frac{2}{3}.$$

These calculations give two more points in the graph, $(0,5.5)$ and $(-5,4\frac{2}{3})$. These points are used in Figure 5.44 to determine how widely the parabola opens.

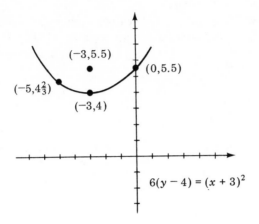

$$6(y - 4) = (x + 3)^2$$

FIGURE 5.44

In Figure 5.45, note that the line from the focus to point P makes the same angle with the tangent line at P as does the horizontal line from P. This is the reason for the parabolic shape of a headlamp. If the bulb is located at the focus, then light rays from the bulb bounce off the reflector in parallel lines, giving a concentrated beam. The principle from physics involved is that the angle of incidence of light rays is equal to the angle of refraction.

Other applications of parabolas will be considered in Section 5.10.

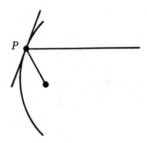

FIGURE 5.45

EXERCISES

1. Locate the vertex, focus, and directrix and graph the following parabolas:

(a)† $8(y - 3) = (x - 2)^2$ (b) $-16(y - 1) = (x - 2)^2$
(c)† $-4(y + 2) = (x - 1)^2$ (d) $4(y - 3) = (x + 1)^2$
(e)† $6(y + 3) = (x + 5)^2$ (f) $-3(y - 1.5) = (x + 3)^2$.

2. Determine the equation of the parabola with:

(a)† Vertex at $(1,-1)$ and directrix $y = 2$.
(b) Vertex at $(2,2)$ and directrix $y = 0$.
(c)† Vertex at $(1,2)$ and directrix $y = 0$.
(d) Vertex at $(-2,3)$ and focus at $(-2,-2)$.
(e)† Focus at $(-3,-2)$ and directrix $y = 2$.
(f) Focus at $(4,0)$ and directrix $y = -1$.
(g)† Vertex at $(0,0)$, $p = \frac{1}{4}$, and opening upward.
(h) Vertex at $(0,0)$, $p = \frac{1}{4}$, and opening downward.

5.9 QUADRATIC EQUATIONS

You may have noticed that the equation of every parabola considered in the last section involved y to the first power only and x to the second power. With some algebraic manipulation, each of those equations could be simplified to the form $y = Ax^2 + Bx + C$, where A, B, and C are some fixed numbers. Such equations are called quadratic equations.

Formally, any equation that can be written in the form

$$y = Ax^2 + Bx + C \quad (A \neq 0)$$

is said to be an equation quadratic in x (or for the sake of simplicity, a *quadratic equation*).

Examples of quadratic equations are:

$$2x^2 - 3x + y = 0.6$$

which can be written as $y = -2x^2 + 3x + 0.6$; $A = -2$, $B = 3$, and $C = 0.6$.

$$y + 3 = x^2$$

which can be written as $y = x^2 - 3$; $A = 1$, $B = 0$, and $C = -3$.

$$2y = 3x^2 + 4x - 1$$

which can be written as $y = \frac{3}{2}x^2 + 2x - \frac{1}{2}$; $A = \frac{3}{2}$, $B = 2$, and $C = -\frac{1}{2}$.

$$y = x^2.$$

What are A, B, and C in this last example?

The important thing to note about the form of an equation quadratic in x is that y always appears to the first power and x always appears to the second power, regardless of whether or not the first power of x appears.

We indicated that the equation of every parabola was a quadratic equation. The converse is also true—the graph of every quadratic equation is a parabola. The procedure whereby a quadratic equation is put into one of our standard forms of a parabola is called *completing the square in x*. This procedure is explained below; for the purpose of illustration, the various steps are performed on the equation $y = 4x^2 - 16x + 20$.

Starting with the equation

$$y = Ax^2 + Bx + C \qquad\qquad y = 4x^2 - 16x + 20$$

1. Divide both sides of the equation by the coefficient of x^2.

1. Dividing both sides of the equation by 4 gives

$$\frac{y}{4} = x^2 - 4x + 5.$$

2. Subtract the constant term from each side of the resulting equation.

2. Subtracting 5 from each side gives $\frac{y}{4} - 5 = x^2 - 4x$.

3. To each side of the resulting equation add the square of one-half the coefficient of x.

3. The square of one-half the coefficient of x is $(-\frac{4}{2})^2 = (-2)^2 = 4$, which is added to each side of the equation to give

$$\frac{y}{4} - 1 = x^2 - 4x + 4.$$

4. The right side can then be factored into the form $(x - h)^2$. (*Note:* $-h$ will always be the quantity which was squared in Step 3.)

4. $\frac{y}{4} - 1 = (x - 2)^2$.

5. The left side can be put into the form $\pm 4p(y - k)$ by factoring the coefficient of y from *both* terms on the left.

5. $\frac{1}{4}(y - 4) = (x - 2)^2$
This is the equation of a parabola with vertex at $(2,4)$ and opening upward. Because $4p = \frac{1}{4}, p = \frac{1}{16}$.

EXAMPLE 1.

Graph the equation $-2y = x^2 - 6x + 11$.

First, the equation must be written in one of the standard forms of a parabola by completing the square in x. The steps will be numbered to conform with the above description.

1. The coefficient of x^2 is 1; so this step can be omitted.
2. $-2y - 11 = x^2 - 6x$

3. $(-\frac{6}{2})^2 = (-3)^2 = 9$
    $-2y - 2 = x^2 - 6x + 9$
4. $-2y - 2 = (x - 3)^2$
5. $-2(y + 1) = (x - 3)^2$

This is the equation of a parabola with vertex at $(3,-1)$ and opening downward. Because $4p = 2$, $p = \frac{1}{2}$ (Fig. 5.46).

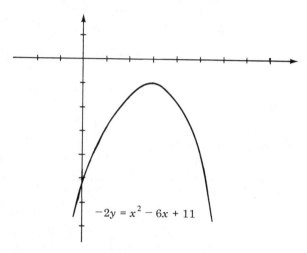

$$-2y = x^2 - 6x + 11$$

FIGURE 5.46

EXAMPLE 2.

Without graphing the equation, determine the coordinates of the vertex of the parabola with equation

$$2y = 3x^2 + 6x + 3.$$

The equation must be put into one of the standard forms by completing the square. The steps are again numbered in conformity with the above description.

1. $\frac{2}{3}y = x^2 + 2x + 1$
2. $\frac{2}{3}y - 1 = x^2 + 2x$
3. $(\frac{2}{2})^2 = 1^2 = 1$
    $\frac{2}{3}y = x^2 + 2x + 1$
4. $\frac{2}{3}y = (x + 1)^2$
5. $\frac{2}{3}(y - 0) = (x + 1)^2$

The vertex is therefore at $(-1,0)$.

EXERCISES

1. Graph the following equations:

   (a)† $y = 4x^2 + 24x + 37$

   (b) $y = -2x^2 + 4x + 1$

   (c)† $y = \dfrac{-x^2}{8} + 4x - 35$

   (d) $y = \dfrac{x^2}{4} - \dfrac{x}{2} + \dfrac{1}{4}$

   (e)† $y = 4x^2 + 8x + 4$

   (f) $y = 2 - x^2$

   (g)† $y = -x^2 - 5x - 4.75$

   (h) $y = x^2 + 0.6x + 0.09$

2. Without graphing the equations, determine the vertex of each of the following parabolas:

   (a)† $3y = x^2 + 2x + 7$

   (b) $y = x^2 - 1$

   (c)† $y = -x^2 + 10x - 29$

   (d) $y = -x^2 + 8x - 16$

   (e)† $y = 8x^2$

   (f) $x^2 + 2y + 2\sqrt{3}x = -1$

*3. Recall that one only has to know the coordinates of two points on a line to determine the equation of the whole line. In the case of a parabola, one has to know only the coordinates of three points on the parabola to determine its equation. The procedure is outlined as follows:

   Joan with her clay pots, in the illustration at the beginning of this chapter, determined the coordinates of three points: $(10, 135)$, $(17, 44)$, and $(25, 60)$, where now the cost is given in cents rather than in dollars. The coordinates of these points will satisfy the quadratic equation $y = Ax^2 + Bx + C$, the graph of which is the parabola containing these three points. For example, substituting the coordinates of the first point for x and y gives

   $$135 = 100A + 10B + C.$$

   (a) Using the coordinates of the two remaining points, determine two more equations in A, B, and C.

   (b) Solve the system made up of the above three equations for the values of A, B, and C. (See Exercise 6, Section 5.6.)

   (c)† Substitute the values of A, B, and C into $y = Ax^2 + Bx + C$ to get the equation of the parabola.

5.10 QUADRATIC EQUATIONS AS MATHEMATICAL MODELS

When the mathematical model for a physical problem is a quadratic equation, the corresponding geometric model is a parabola. This type of model is particularly advantageous because of the role played by the vertex.

In the case of a parabola opening downward, its vertex (h,k) is the highest point on the graph; k is therefore the largest value acquired by y, and h is the value of x at which this largest value is acquired. In the case of a parabola opening upward, its vertex (h,k) is the lowest point on the graph; k is therefore the smallest value acquired by y, and h is the value of x at which this smallest value is acquired.

This information can often be used to great advantage, as the following examples illustrate.

EXAMPLE 1.

A company ships its products in cardboard boxes 2 inches deep. The cost of these containers is based upon the perimeter of the base of the boxes. The company decides that its most economical size is a box with a perimeter of 36 inches. What size box meeting these specifications will have the maximum volume?

The volume of a rectangular container is length times width times depth. Let x denote the width in inches; for the perimeter to be 36 inches, the length must be $18 - x$ inches. The volume (y) is therefore

$$y = 2(18 - x)x = -2x^2 + 36x$$

cubic inches.

Completing the square in x gives

$$-\tfrac{1}{2}(y - 162) = (x - 9)^2,$$

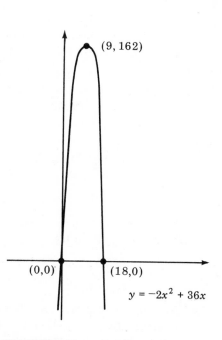

which is a parabola opening downward with vertex $(9, 162)$. The maximum volume is 162 cubic inches and is attained when the width is 9 inches. Note that those points of the graph in Figure 5.47 with x-coordinates less than zero or greater than 18 are irrelevant for this particular situation. Not only must the width be positive $(x > 0)$, but it must also be less than 18 inches $(x < 18)$ because the sum of the length and width can be only 18 inches.

FIGURE 5.47

EXAMPLE 2.

The cost to the Styl Clothing Company for manufacturing shirts of a particular type is given by

$$y = x^2 - 100x + 2,750$$

where x is the number of shirts produced, and y is the cost, in cents, per shirt. The company decides that the number of shirts that it will manufacture is the number at which the cost is minimal. How many shirts should the company produce? What will be the total cost? What will be the cost per shirt?

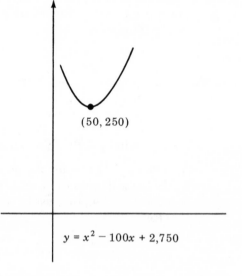

(50, 250)

$y = x^2 - 100x + 2,750$

Completing the square in the cost equation gives

$$y - 250 = (x - 50)^2,$$

whose graph (Fig. 5.48) is a parabola opening upward with vertex at $(50, 250)$. The minimal cost per shirt

FIGURE 5.48

will occur if the company produces 50 shirts; at this level, the per shirt cost would be $2.50. Total cost: $2.50 X 50 = $125.00.

In the graph of the equation, only those points with integral x-coordinates greater than or equal to zero are relevant. Partial shirts or negative numbers of shirts are not manufactured!

EXAMPLE 3.

For Joan, in the illustration at the beginning of this chapter, the fuel cost is given by

$$y = x^2 - 40x + 435,$$

where x is the lot size, and y is the fuel cost, in cents, per pot. (See Exercise 3, Section 5.9.)

Completing the square in x gives

$$(y - 35) = (x - 20)^2.$$

The graph of this equation is a parabola which opens upward and has its vertex at $(20, 35)$. The minimal cost is therefore 35¢, which is obtained when the lot size is 20.

EXERCISES

1.[†] The total annual profit (y), in dollars, of an automobile accessories company from the manufacture and sale of x rear-view mirrors is given by

$$y = \frac{-x^2}{400} + 2x - 75.$$

   (a) How many mirrors must the company sell per year to achieve maximum profit?
   (b) What is the maximum annual profit that the company can achieve from the sale of the mirrors?
   (c) What is the profit per mirror when the maximum is achieved?
   (d) Graph the profit equation.
   (e) Which points on the graph are relevant to this particular situation?

2. (a) What is the maximum income that the airline company of Exercise 13, Section 4.1, can receive from its charter flight?
   (b) How many empty seats are there when the maximum income is attained?
   (c) What is the price per person when the maximum is attained?

3.[†] (a) What is the maximum profit that the farmer of Exercise 14, Section 4.1, can achieve per year?
   (b) How many additional trees must he plant to achieve the maximum profit?
   (c) What is the profit per tree when the maximum is achieved?

4. A large department store estimates that it loses $400 per month in sales in a particular department because of a shortage of personnel. A management consultant advises that if an additional part-time employee were to be hired to work x hours per month in the department, the loss (y) in sales would be

   $$y = 400 + x(x - 20).$$

   How many hours should the part-time employee work to attain minimum loss of sales?

5.[†] What are the dimensions of the largest garden that Pat of Exercise 15, Section 4.1, can fence?

6. What is the maximum total income that Paul of Exercise 16, Section 4.1, can receive per month from his apartment complex?

5.11 SUMMARY

By means of two number lines (the x-axis and the y-axis), the points of a plane can be put into one-to-one correspondence with the set of all ordered pairs of real numbers. The numbers associated with a particular point are called the coordinates of that point. The resulting plane is called a coördinate system and provides the setting for analytic geometry.

The primary concern in analytic geometry is the relationship between an equation and its graph—its graph being the set of all points whose coordinates satisfy the equation. This relationship can be considered from two points of view: given an equation, determine its graph; or given a graph, determine its equation.

Starting with a straight line, we found its equation to be

$$y - y_0 = m(x - x_0)$$

where (x_0, y_0) is a point on the line, and m is its slope, or

$$y = mx + b$$

where m is again the slope, and $(0,b)$ is the point at which the line crosses the y-axis. If a line is parallel to the y-axis, it has no slope and its equation is

$$x = c$$

where c is the x-coordinate of any point on the line.

Each of these three equations can be rewritten as a linear equation

$$Ax + By + C = 0$$

for appropriate values of A, B, and C.

Conversely, starting with a linear equation, one finds that its graph is a straight line. If $B \neq 0$, the equation can be written

$$y = -\frac{A}{B}x - \frac{C}{B},$$

and its graph is the line passing through the point $(0, -C/B)$ with slope $-A/B$. If $B = 0$, the equation can be written

$$x = -C/A,$$

and its graph is a line through the point $(-C/A, 0)$ parallel to the y-axis.

A parabola is defined to be the set of all points that are equidistant from a fixed point, the focus, and a fixed line, the directrix. Starting with the graph of a parabola, we found its equation to be

$$4p(y - k) = (x - h)^2$$

if the parabola opens upward, or

$$-4p(y - k) = (x - h)^2$$

if it opens downward. In each case, (h,k) gives the coordinates of the vertex of the parabola. The number p is the distance from the focus to the vertex.

Each of the two equations for parabolas can be rewritten as a quadratic equation

$$y = Ax^2 + Bx + C$$

for appropriate values of A, B, and C.

Conversely, starting with a quadratic equation

$$y = Ax^2 + Bx + C,$$

we found that its graph is a parabola. By completing the square in x, the quadratic equation can be rewritten in one of the two forms given above, revealing the location of its vertex, the value of p, and the direction in which the parabola opens.

When the mathematical model for a physical problem consists of an equation, the graph of the equation gives a picture of the relationship between the quantities involved. When the graph is a parabola, the coordinates of the vertex give the minimum or maximum value that y assumes along with the value of x at which the minimum or maximum is attained.

FIGURE 5.49
The Saarinen-designed 630-foot stainless steel Arch on the western bank of the Mississippi in St. Louis. The shape of the Arch is a *catenary*, the shape assumed by a chain suspended by its two ends. A cross-section of each leg is an *equilateral triangle*.

This photo illustrates some of the applications of geometry in architecture.

The heights of the boy in the picture form a
geometric progression.

chapter 6

geometric progressions and consumer finance

Can you imagine yourself, as the grand winner of a TV quiz program, being allowed to choose for your prize a check for $100,000 or a checkerboard with the red squares loaded with pennies as shown in Figure 6.1?

FIGURE 6.1

On the first red square will be one penny; on the second red square will be two pennies; on the third, four pennies, on the fourth, eight pennies, and so forth. The distribution of pennies will continue until all thirty-two red squares are filled, each with twice as many pennies as the previous. Which should you choose?

Certainly, $100,000 sounds like a lot of money compared to a checkerboard with a bunch of pennies. But that doubling of pennies might add up pretty fast. As we shall see, you would actually do considerably better by choosing the pennies. Your major problem would be to haul them home—and pay the income tax on them!

Or suppose that your parents started a college fund for you on your first birthday by depositing $200 in a savings account which pays 6% interest. Each subsequent year, they deposited $200. If the interest was allowed to accumulate, how much would be available to you when you started college?

If you emptied your account after the $200 deposit on your eighteenth birthday, you would have about $6,180—quite a bit more than the $3,600 deposited over the years.

Surprising, perhaps, is the fact that the sum of money on the checkerboard and the amount of money in your account can be calculated in exactly the same way. The amounts of money on each square and the amounts of money in your account at the end of each year both form a sequence of numbers called a geometric progression, the subject to be studied in the next section. As such, geometric progressions are the mathematical models of these amounts. One of the topics to be considered is the sum of the numbers in a geometric progression. We will then be in a position to analyze further the two illustrations discussed above.

6.1 GEOMETRIC PROGRESSIONS

Consider the following sequences of numbers:

$1, 3, 9, 27, 81, 243, 729, \ldots$
$2, -4, 8, -16, 32, -64, 128, \ldots$
$64, 32, 16, 8, 4, 2, 1, \frac{1}{2}, \frac{1}{4}, \ldots$

In the first sequence, each number after the first can be obtained from the preceding by multiplying by 3; in the second, by multiplying by -2; and in the third, by multiplying by $\frac{1}{2}$. Sequences of numbers such as these are called geometric progressions.

Formally, a *geometric progression* is a sequence of numbers such that each number in the sequence, after the first, can be obtained from the preceding number by multiplying by some constant. This constant is called the *common ratio* and is usually denoted by the letter r. In the first illustration above, $r = 3$; in the second, $r = -2$; and in the third, $r = \frac{1}{2}$.

The numbers in a geometric progression are called the *terms*. The first term is denoted by a_1, the second by a_2, the third by a_3, and so forth. In the geometric progression

$$\tfrac{1}{125}, \tfrac{1}{25}, \tfrac{1}{5}, 1, 5, 25, 125, \ldots,$$

$a_1 = \tfrac{1}{125}, a_2 = \tfrac{1}{25}, a_3 = \tfrac{1}{5}, a_4 = 1$, and so forth. In the geometric progression giving the number of pennies on the checkerboard squares, $a_1 = 1, a_2 = 2$, $a_3 = 4, a_4 = 8$, and so on. What are the values of r in each of these last two cases?

In terms of the a's and r:

$$a_2 = r a_1$$
$$a_3 = r a_2$$
$$a_4 = r a_3$$
$$a_5 = r a_4$$
$$\cdots$$

That is, each term can be determined from the preceding by multiplying by r, as the definition requires. Or,

$$a_3 = r^2 a_1$$
$$a_4 = r^2 a_2$$
$$a_5 = r^2 a_3$$
$$a_6 = r^2 a_4$$
$$\cdots;$$

that is, each term can be determined from the second preceding by multiplying by r^2.

In terms of a_1 and r, we have

$$a_2 = r a_1$$
$$a_3 = r^2 a_1$$
$$a_4 = r^3 a_1$$
$$a_5 = r^4 a_1$$
$$a_6 = r^5 a_1$$
$$\cdots$$

In general, any term $a_n = r^{n-1} a_1$. This relationship enables one to determine any term in a progression without calculating all of the preceding terms. For example, if $a_1 = 4$ and $r = 3$, the seventh term is

$$a_7 = 3^6 \times 4 = 2{,}916.$$

On the checkerboard, the number of pennies on the 32nd red square would be

$$a_{32} = 2^{31} \times 1 = 2{,}147{,}483{,}648.$$

EXAMPLE 1.

Determine the first five terms of the geometric progression in which $a_2 = 36$ and $a_3 = 108$.

Because $a_3 = ra_2$, we have $108 = r(36)$, from which it follows that $r = 3$. Once the value of r is known, it is easy to determine the three missing terms.

$$a_4 = 3a_3 = 324$$
$$a_5 = 3a_4 = 972$$
$$a_1 = a_2/3 = 12.$$

The first five terms are

$$12, 36, 108, 324, 972.$$

EXAMPLE 2.

Determine the first five terms of the geometric progression in which $a_1 = 2$ and $a_3 = 18$.

Because $a_3 = r^2 a_1$, we have $18 = r^2(2)$, so that $r^2 = 9$. Consequently, r could be 3 or -3. Each case must be treated separately. If $r = 3$,

$$a_1 = 2$$
$$a_2 = 6$$
$$a_3 = 18$$
$$a_4 = 54$$
$$a_5 = 162.$$

If $r = -3$,

$$a_1 = 2$$
$$a_2 = -6$$
$$a_3 = 18$$
$$a_4 = -54$$
$$a_5 = 162.$$

EXAMPLE 3.

Can $a_1 = 2$, $a_2 = 4$, and $a_3 = 1$ be the first three terms of a geometric progression?

Comparing a_1 and a_2 determines that r must have the value $\frac{4}{2} = 2$.

Comparing a_2 and a_3 determines that r must have the value $\frac{1}{4}$. Consequently, no geometric progression can begin with the terms 2, 4, 1, because the value of r must be constant in any one progression.

EXAMPLE 4.

Ed has a ton of gravel, which he has to spread himself, delivered for his driveway. The chore does not excite him, so he decides to spread one-half the gravel the first day, and on subsequent days about one-third of the previous day. He figures that his ambition will run out by about the fifth day, but by then the job will be completed anyway. Will it?

The daily amounts that he spreads form a geometric progression with $a_1 = \frac{1}{2}$ and $r = \frac{1}{3}$:

$$\frac{1}{2}, \frac{1}{6}, \frac{1}{18}, \frac{1}{54}, \frac{1}{162}, \cdots .$$

If we add the first five terms, we get

$$\frac{1}{2} + \frac{1}{6} + \frac{1}{18} + \frac{1}{54} + \frac{1}{162} = \frac{121}{162}$$

or about $\frac{3}{4}$ tons. The job will not be completed!

EXAMPLE 5.

If Ed in Example 4 decides instead to spread one-third of the remaining amount of gravel each day, do the daily amounts form a geometric progression?

The amount of gravel spread the first day is $\frac{1}{3}$ ton; $a_1 = \frac{1}{3}$. This leaves $\frac{2}{3}$ ton, $\frac{1}{3}$ of which he spreads the second day; $a_2 = \frac{1}{3} \times \frac{2}{3} = \frac{2}{9}$. There remains $\frac{2}{3} - \frac{2}{9}$ or $\frac{4}{9}$ ton, so that on the third day the number of tons spread is $a_3 = \frac{1}{3} \times \frac{4}{9} = \frac{4}{27}$.

Comparing a_1, a_2, and a_3, it looks as though the amounts do form a geometric progression with $r = \frac{2}{3}$, but additional terms must be examined. In Exercise 4 of this section, you are asked to determine a_4 and a_5, and to finish this example.

EXAMPLE 6.

Mark is offered a job with a starting salary of $1,000 per month and is told that the average yearly raise is about 6%. On this basis, what will be his monthly salary during the fifth year?

His monthly salaries form a geometric progression with $a_1 = 1,000$ and $r = 1.06$. His salary for the fifth year will be, in dollars,

$$a_5 = (1.06)^4 \times 1,000 = 1,262.48.$$

___

Quantities that are positive and that form a geometric progression with a common ratio greater than 1 are said to *increase geometrically*. The ratio need not be much bigger than 1 for these quantities to increase very rapidly. Recall that on the checkerboard, with $r = 2$, the number of pennies on the thirty-second square was 2,147,483,648.

On the other hand, quantities that are positive and that form a geo-metric progression with r less than 1 are said to *decrease geometrically*. These quantities generally tend to decrease very rapidly. In Figure 6.2, the various squares are obtained by dividing the next larger square into fourths. The areas of the different size squares form a geometric progres-sion with $r = \frac{1}{4}$. After only six different sizes, the areas of the smallest square is $\frac{1}{1024}$ times that of the largest.

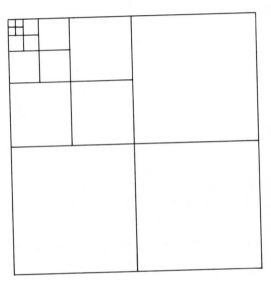

FIGURE 6.2

EXERCISES

1. Which of the following can be the first three terms of a geometric progression?

    (a)† $5, 15, 45$

    (b) $5, 1, \frac{1}{5}$

    (c)† $0.001, 0.005, 0.025$

    (d) $-1, -\frac{1}{2}, -\frac{1}{4}$

    (e)† $-5, -15, 45$

    (f) $0.1, 1, 100$

2. Without calculating any other terms, find

    (a)† a_3 if $a_1 = 4$ and $r = 5$.

    (b) a_4 if $a_1 = 3$ and $r = -2$.

    (c)† a_7 if $a_1 = 2$ and $r = -1$.

    (d) a_6 if $a_1 = -5$ and $r = \frac{1}{2}$.

    (e)† a_5 if $a_1 = -2$ and $r = -\frac{1}{3}$.

    (f) a_4 if $a_1 = 7$ and $r = 0.1$.

3. Determine the first five terms of the geometric progression(s) for which

   (a)† $a_1 = 3$ and $r = 4$.
   (c)† $a_1 = 1$ and $r = 0.01$.
   (e)† $a_2 = \frac{1}{9}$ and $a_3 = \frac{1}{3}$.
   (g)† $a_2 = 1.5$ and $a_4 = 13.5$.

   (b) $a_1 = -16$ and $r = \frac{1}{4}$.
   (d) $a_1 = -2$ and $a_2 = 4$.
   (f) $a_1 = 15$ and $a_3 = 375$.

4. Determine whether or not the amount of gravel spread the first five days in Example 5 of this section form a geometric progression.

5.† The number of a certain type of bacteria in an organism decreases by about one-third every hour for the first four hours after an antibiotic is administered. Determine the number of bacteria present at the end of the fourth hour after the antibiotic is administered in a specimen originally containing 9,000. (*Hint:* What should r be to count the number present?)

6. If it takes $15,000 this year to maintain a family of four at a reasonable standard of living, and if the rate of inflation is 6% a year for the next three years, how much money will be required to maintain the same family at the same standard of living three years hence? Determine the correct amount by finding the fourth term in an appropriate geometric progression. (*Hint:* $r = 1.06$.)

7.† A poor "wino" has devised a method to make a quart of wine last longer. At each nip, he drinks half the quart and then replaces the amount he drank with water; at the fourth nip he drinks the whole quart. How much of the wine has he drunk at the end of the third nip?

*8. A water lily that doubles its size every day is planted in a pond. If it fills the pond in ten days, how many days would it take two such water lilies to fill the pond?

It was pointed out earlier that one of our major objectives is to find a method for determining the sum of the numbers in a geometric progression. One could, of course, calculate the desired terms and add them up—but this method becomes cumbersome when the number of terms is very large. There is a shorter method which requires only the values of a_1 and of r.

Suppose we wanted to add the first n terms in a geometric progression, and we denote their sum by S_n:

$$S_n = a_1 + a_2 + a_3 + \cdots + a_{n-1} + a_n$$

or, in terms of a_1 and r,

$$S_n = a_1 + a_1 r + a_1 r^2 + \cdots + a_1 r^{n-2} + a_1 r^{n-1}.$$

If we multiply both sides of this last equation by r, we get

$$rS_n = a_1 r + a_1 r^2 + a_1 r^3 + \cdots + a_1 r^{n-1} + a_1 r^n.$$

Subtracting corresponding sides of these last two equations gives

$$
\begin{array}{l}
S_n = a_1 + a_1 r + a_1 r^2 + \cdots + a_1 r^{n-1} \\
- rS_n = \quad\ -a_1 r - a_1 r^2 - \cdots - a_1 r^{n-1} - a_1 r^n \\
\hline
S_n - rS_n = a_1 - a_1 r^n.
\end{array}
$$

Each side of this last equation can be factored to give

$$S_n(1-r) = a_1(1-r^n)$$

or

$$S_n = \frac{a_1(1-r^n)}{1-r} \qquad\qquad (6.1)$$

Equation (6.1) gives a method of determining S_n using only the values of a_1 and of r (except for $r = 1$).

EXAMPLE 7.

Calculate the sum of the first six terms in the geometric progression with $a_1 = 2$ and $r = 4$.

In this case, n, the number of terms to be added is 6. We therefore want to calculate

$$S_6 = \frac{2(1-4^6)}{1-4} = \frac{2(1-4{,}096)}{-3} = 2{,}730.$$

EXAMPLE 8.

Calculate the sum of the first seven terms in the geometric progression in which $a_2 = 3$ and $a_3 = -6$.

Comparing a_2 and a_3, we see that $r = -2$. Also, $a_1 = a_2/r = -\frac{3}{2} = -1.5$. Because $n = 7$, we calculate

$$S_7 = \frac{-1.5[1-(-2)^7]}{1-(-2)} = \frac{-1.5[1-(-128)]}{1+2} = -64.5.$$

EXAMPLE 9.

Determine the amount of money on the checkerboard in the illustration at the beginning of this chapter.

In this case, $n = 32$, $a_1 = 1$, and $r = 2$.

$$S_{32} = \frac{1(1-2^{32})}{1-2} = \frac{1-4{,}294{,}967{,}296}{-1} = 4{,}294{,}967{,}295.$$

Consequently, there would be $42,949,672.96. This is more than anyone is ever likely to be giving away—except perhaps for the Federal Government!

Note the procedure discussed in Chapter 4 for the application of mathematics at work in this last example. The physical question was, "How much money is on the checkerboard?" The mathematical model of the amount of pennies was a geometric progression. The mathematical question was, "What is the sum of the first thirty-two terms in this progression?" The interpretation of the mathematical solution in terms of dollars and cents gave the answer to the original physical question.

EXERCISES

1. Find the sum of the indicated number of terms using Equation (6.1):
   (a)† the first 5 terms, $a_1 = 1, r = 2$.
   (b) the first 7 terms, $a_1 = -3, r = 2$.
   (c)† the first 4 terms, $a_1 = 6, r = \frac{1}{3}$.
   (d) the first 5 terms, $a_1 = -2, a_2 = -1$.
   (e)† the first 3 terms, $a_2 = 1, a_3 = -4$.
   (f) the first 1,000 terms, $a_1 = 0, r = 5$.

2. Use Equation (6.1) twice to find the sum of the seventh, eighth, and ninth terms of the geometric progression in which $a_1 = 2$ and $r = 3$.

3. Use Equation (6.1) to determine the amount of gravel that Ed of Example 4 of this section spread the first five days.

*4. Use Equation (6.1) to calculate the average (mean) monthly salary that Mark of Example 6 of this section will make during the first five years.

*5.† Determine an expression for the sum of the first n terms of a geometric progression if $r = 1$. (Note that Equation (6.1) does not apply if $r = 1$.)

6.2 SIMPLE INTEREST

This section begins the second part of the subject matter of this chapter, the mathematics of finance—a quite technical subject when considered in full detail. Our approach will be to avoid as much of the technicality as possible and to consider the subject from the point of view of the consumer.

The object is to enable you to gain some insight into the workings of finance—to feel more comfortable with a topic none of us can entirely avoid given our present economic system.

Applications of geometric progressions do not occur in our discussion of simple interest. However, they do appear again in the following section on compound interest.

The interest involved in any money-lending situation has two aspects—there are two sides of the coin, so to speak. On the side of the borrower, interest is payment made to someone else for the use of his or her money. In some situations, the money itself may be passed from one person to another, as when a bank loan is made, but it need not be. When a purchase is made by a credit card or on an installment plan, the buyer is using the money that the seller has put into the item purchased.

From the point of view of the lender, interest is payment received for the use of his or her money. This is true whether the lender is a savings and loan association financing a new home or an individual putting his or her money into a savings account so that the bank can use the money for its commercial purposes.

Simple interest is determined only by the amount of money borrowed or loaned, the rate of interest, and the duration of the loan. If you were to borrow $150 from your bank at 6% per year for two years, the interest you would pay would be

$$150 \times 0.06 \times 2 = 18,$$

or $18.00. This $18.00 would be simple interest.

The $150 in the above transaction is the *principal*, the 6% per year is the interest *rate*, and the loan duration of two years is called the *time*. With this terminology, the interest owed at the end of two years is given by

Principal × Rate × Time

or, with obvious notation,

$$I = PRT. \qquad (6.2)$$

Simple interest is always calculated using Equation (6.2).

At the end of two years, the total amount you would have to pay the bank would be $168, the principal plus the interest,

$$P + I$$

or, from Equation (6.2),

$$P + PRT = P(1 + RT).$$

This total sum to be repaid is called the *amount* and, for simple interest, is given by

$$A = P(1 + RT) \qquad (6.3)$$

Should your loan have been arranged in such a way that, during the second year, you would pay interest on the interest accumulated during the first year, the interest on the loan would have been slightly higher:

For the first year: $I = 150 \times 0.06 \times 1 = 9.00$
For the second year: $I = 159 \times 0.06 \times 1 = 9.44$

Total interest: $\overline{18.44}$

When interest is paid on interest, as in this last case, the result is compound interest. This section treats only simple interest and considers various situations in which simple interest is applicable.

Note that both the interest rate and the duration of a loan involve a time factor. Both are expressed in terms of days, months, or years. When using Equations (6.2) or Equation (6.3), *both must be expressed in terms of the same unit of time.*

EXAMPLE 1.

Steven borrowed $200 from his credit union for two months at 9% per year. Find the amount of interest that he will have to pay.

Because two months equals $\frac{1}{6}$ year, the interest he will have to pay is

$$200 \times 0.09 \times \tfrac{1}{6} = 3.00.$$

If we had chosen to do the calculations in terms of months, the interest rate must be expressed as $\frac{9}{12}$, or $\frac{3}{4}$ % per month. The amount of interest will be the same:

$$200 \times 0.0075 \times 2 = 3.00.$$

(Note that $\frac{3}{4}\% = \frac{3}{4} \times \frac{1}{100} = 0.0075$.)

Throughout this text, *all interest rates will be understood to be annual rates, unless otherwise indicated.*

Purchases made by credit cards generally involve simple interest. When payments extend over several months, the total interest paid requires separate calculations for each month.

EXAMPLE 2.

Pat uses her credit card to buy a stereo from a department store for $325. She intends to pay $100 a month for the first two months, and the balance the third month. The department store does not charge interest on payments made the first thirty days, but does charge $1\frac{1}{2}$ % per month thereafter. Find the total amount that she pays for the stereo.

1st Month

Interest: None
Payment: 100.00
Balance: $325.00 - 100.00 = 225.00$

2nd Month

Interest: $225.00 \times 0.015 \times 1 = 3.38$
Payment: 100.00
Balance: $225.00 + 3.38 - 100.00 = 128.38$

3rd Month

Interest: $128.38 \times 0.015 \times 1 = 1.93$
Payment: $128.38 + 1.93 = 130.31$
Balance: $128.38 + 1.93 - 130.31 = 0$

Total Payments: 100.00
                    100.00
                    130.31
                    ‾‾‾‾‾‾
                    330.31

The total amount (excluding any sales tax) Pat paid for the stereo was $330.31.

Note that the interest each month, after the first month, is calculated on the previous month's balance and is added to the current month's balance. Note also that the $1\frac{1}{2}$ % monthly interest rate is equivalent to 18% annually.

Another type of consumer loan is the installment loan with add-on interest. On this type of loan, the amount (principal plus interest) is divided evenly over a fixed number of payments.

EXAMPLE 3.

Suppose that Pat of Example 2 bought the stereo on an installment plan with add-on interest of 18%, and with payments due at the end of each of the next three months. How much would each installment amount to?

The interest is still simple interest:

$$325 \times 0.18 \times \tfrac{1}{4} = 14.63.$$

This charge is added to the cost of the stereo and the total amount is divided by 3 (the number of payments).

        325.00
         14.63
      ‾‾‾‾‾‾‾‾‾
    3 | 339.63
        113.21

Each payment would be $113.21. (Compare the total cost here with that of Example 2; both have an annual rate of 18%.)

In an installment loan with add-on interest, the consumer pays interest on the principal for the entire length of the loan, even though the money involved in the first payments is used only a short time. This causes the true annual rate to be higher than the declared add-on rate. The Truth-in-Lending law, to be considered later in this chapter, requires the lender to specify the true annual rate in all consumer loan transactions.

Home mortgage loans are not the same. The consumer pays interest on each part of the principal only during the time that part is actually used. Each month, the interest on the unpaid principal is paid first; the balance of the monthly payment is used to reduce the principal. This procedure is illustrated in Example 4.

EXAMPLE 4.

A couple buys a house for $44,000, makes a 25% down payment, and finances the balance with a 20-year home mortgage at $8\frac{1}{2}\%$. Their monthly payments are $286. Determine how much of the first two payments will be for interest and how much will reduce the principal.

The principal will be the cost of the house less the down payment:

$$44,000 - 11,000 = 33,000.$$

1st Month

Interest: $33,000 \times 0.085 \times \frac{1}{12} = 233.75$
Principal: $286.00 - 233.75 = 52.25$
Balance: $33,000 - 52.25 = 32,947.75$

2nd Month

Interest: $32,947.75 \times 0.085 \times \frac{1}{12} = 233.38$
Principal: $286.00 - 233.38 = 52.62$
Balance: $32,947.75 - 52.62 = 32,895.13$

You should note carefully in Example 4 the manner in which the interest is calculated and paid each month. After the interest is paid, the rest of the monthly payment is used to reduce the principal. The manner in which a home mortgage loan is handled will be important in Section 6.4, where the Truth-in-Lending law is considered.

EXERCISES

1.† Determine the amount of interest earned on $5,000 invested for two years at 6%.

2. Ed borrows $350 from his bank for 90 days at 8%. How much interest will he have to pay? What amount will he owe the bank at the end of the 90 days? (Consider a year to have 360 days.)

3.† Mark deposits $200 in his credit union account, which pays $5\frac{1}{2}\%$ interest. Determine the amount in his account four months later when the interest is added.

4. Suppose that your checking account service charge runs about $2.50 each month. If you maintain a minimum balance of $200 each month, there would be no service charge. Would it be to your advantage to move $200 from your savings account, which pays 5%, to your checking account to maintain the minimum balance?

5.† Joan bought a $250 bicycle with her credit card from a department store with the same charge plan as that of Example 2 of this section. She plans to pay $75 a month until the bicycle is paid for. Determine the total amount of interest she will have to pay.

6. On his trip to the Republican convention, Dan charged a total of $225 in restaurant and hotel bills on his bank card. The bank has the same charge plan as the department store of Example 2 of this section. Find the total interest he will have to pay if he expects to make payments of $100 a month.

7.† An automobile dealer allows Laura a $1,100 trade-in on a new car costing $3,600. She finances the balance with an installment loan charging 7% add-on interest. Determine how much each of her 36 monthly payments will be.

8. For $25 down and 24 monthly payments of $100 each, you can buy a motorcycle costing $1,825. Find the rate of add-on interest you will be paying.

9.† Harry and Mary bought a $25,000 house, paying 20% down and financing the balance with a 9% mortgage loan. Their monthly payments are $178. Determine how much of their first two payments will be for interest and how much will reduce the principal.

When a bank calculates the amount of interest based on the final amount of the loan rather than the actual amount of money borrowed, the loan is said to be *discounted*. If an individual takes out a loan of $300 for two years discounted at 7%, the amount of money he or she actually receives is

$$300 - (300 \times 0.07 \times 2) = 258$$

dollars; the interest is subtracted from the principal when the loan is made. The effect is that the individual is paying a rate of interest higher than 7% on the $258 received.

EXAMPLE 5.

Jack borrowed $250 for three months discounted at 8%. What rate of interest is he paying on the amount of money that he receives?

The amount of money that he received is

$$250 - (250 \times 0.08 \times \tfrac{1}{4}) = 245$$

dollars, for which he is paying $5.00 interest. We want to determine what interest rate this $5.00 represents relative to the $245 he receives.

From Equation (6.2), we get for rate

$$R = \frac{I}{PT}. \tag{6.4}$$

Using the appropriate values in Equation (6.4), Jack's interest rate is

$$R = \frac{5}{245 \times \tfrac{1}{4}} = 0.082,$$

or 8.2% per year.

EXAMPLE 6.

Eric is buying a piece of land for which he can either pay $6,000 now or $6,500 in three years. He has the $6,000 invested at $5\tfrac{1}{2}\%$. Should he use the $6,000 to pay for the land now or should he wait?

The interest that he would earn on the $6,000 during the three years would be

$$6,000 \times 0.055 \times 3 = 990$$

dollars. Because his $6,000 would be worth $6,990 in three years, he would be ahead $490 at that time by waiting.

An alternative method of deciding the question asked in this last example is to find the sum of money which, when invested now at $5\tfrac{1}{2}\%$, would yield an amount of $6,500 three years hence. This sum, called the *present value* of $6,500, is easy to determine.

From Equation (6.3), which gives the amount in terms of principal, rate, and time, we get

$$P = \frac{A}{1 + RT} \tag{6.5}$$

Equation (6.5) gives the present value of any amount for simple interest investments. In Example 6

$$P = \frac{6,500}{1 + (0.055 \times 3)} = \frac{6,500}{1.165} = 5,579.40,$$

which also indicates that Eric should wait. If $5,579.40 is invested now at $5\frac{1}{2}\%$, it will amount to $6,500 in three years; obviously, $6,000 at the same rate will amount to more.

EXERCISES

1. How does Equation (6.4) follow from Equation (6.2)?

2. How does Equation (6.5) follow from Equation (6.3)?

3.† Catherine borrowed $1,200 for one year discounted at 10%.
   (a) How much money did she receive when the loan was made?
   (b) How much interest is she paying for the money she actually received?

4. Sally bought a ninety-day (three-month) $10,000 U.S. Treasury bill discounted at 8%; that is, she paid $9,800 for the bill, which the Treasury Department will redeem in 90 days for $10,000. What rate of interest will she be making on her $9,800?

5.† Kay buys five $1,000 corporate bonds, which every six months pay interest at the rate of 18% per year. Determine the amount of the interest check that she receives every six months.

6. Kay of Exercise 5 above thought that she could make better use of her money, so she sold her bonds to Susan for $925 each. Susan now receives the semiannual payments. What rate of interest on her investment is Susan receiving from these payments?

7.† After Susan in Exercise 6 above has the bonds for ten years, the corporation redeems them (buys them back) for $1,000 each. Find the sum of all her semiannual payments and her profit from the sale on the bonds. Use this sum to determine the annual rate of return on her investment.

8. Gene can settle a debt by either paying $200 now or $208 in six months. He has the money invested at $5\frac{1}{2}\%$. Which is the better plan for him?

9.† How much money invested now at 6% will amount to $450 in two years?

6.3 COMPOUND INTEREST

It is probably accurate to say that most financial transactions that involve an extended period of time are based on compound interest. In any case, it is an integral element in our financial system. Indicative of its importance is the following statement by Leonard S. Silk. "Explaining compound interest—for a bank or a nation—is the central problem of economics."[1]

We had a glimpse into the determination of compound interest in the last section; before examining this topic in more detail, let us take a look at the calculations over a longer period of time. In particular, we will determine the interest for four years on $1,000 invested at 5% compounded each year.

1st Year

Interest: $1,000 \times 0.05 \times 1 = 50$
Compound Amount: $1,000 + 50 = 1,050$

2nd Year

Interest: $1,050 \times 0.05 \times 1 = 52.50$
Compound Amount: $1,050.00 + 52.50 = 1,102.50$

3rd Year

Interest: $1,102.50 \times 0.05 \times 1 = 55.13$
Compound Amount: $1,102.50 + 55.13 = 1,157.63$

4th Year

Interest: $1,157.63 \times 0.05 \times 1 = 57.88$
Compound Amount: $1,157.63 + 57.88 = 1,215.51$

Note that compound interest is simple interest computed again and again. The difference between compound and simple interest is that, after the first period, compound interest is based on the interest from the previous periods as well as on the original principal; simple interest is based only on the original principal.

To compare the effects of compound and simple interest, consider Table 6.1, which gives the balance in an imaginary account at the end of a number of years for both compound and simple interest.

Table 6.1 indicates that interest on interest, even though it may only amount to a few dollars in the beginning, can soon become a significant amount.

The year-by-year computation to determine the compound amount, as in the illustration above, soon becomes very tedious. So one looks for an easier method. Consider the general case where P dollars are invested

---

[1] Leonard S. Silk, *The Research Revolution* (New York: McGraw Hill Book Co., Inc., 1960), p. 38.

TABLE 6.1   Account Balance on $1,000 Invested at 5%

At end of Year	Compound Interest	Simple Interest
1	1,050.00	1,050.00
2	1,102.50	1,100.00
3	1,157.63	1,150.00
4	1,215.51	1,200.00
5	1,276.28	1,250.00
6	1,340.10	1,300.00
7	1,407.10	1,350.00
8	1,477.46	1,400.00
9	1,551.33	1,450.00
10	1,628.89	1,500.00
15	2,078.93	1,750.00
20	2,653.30	2,000.00

(or borrowed) at a rate of i interest per year. The computations are as follows.

1st Year
Interest: $P \times i = Pi$ (Time = 1, so it is omitted.)
Compound Amount: $P + Pi = P(1 + i)$

2nd Year
Interest: $P(1 + i) \times i = P(1 + i)i$
Compound Amount: $P(1 + i) + P(1 + i)i = P(1 + i)[1 + i] = P(1 + i)^2$

3rd Year
Interest: $P(1 + i)^2 \times i = P(1 + i)^2 i$
Compound Amount: $P(1 + i)^2 + P(1 + i)^2 i = P(1 + i)^2[1 + i] = P(1 + i)^3$.

In each case, the compound amount at the end of the nth year is $P(1 + i)^n$. Continued calculations would verify that the same expression gives the compound amount for any number of years. Consequently, if we denote the compound amount by C, then C is given by the equation

$$C = P(1 + i)^n. \tag{6.6}$$

EXAMPLE 1.

Find the compound amount of $1,000 invested for four years at 5%.

In this case, $P = 1,000$, $i = 0.05$, and $n = 4$. From Equation (6.6), we get

$$C = 1,000(1 + 0.05)^4 = 1,000(1.05)^4 = 1,000(1.21551) = 1,215.51.$$

The compound amount is $1,215.51. (Compare this result with the illustration at the beginning of this section.)

Equation (6.6) was derived on the assumption that i was the annual rate and that the interest was compounded annually. Although i is normally stated on an annual basis, the interest is usually compounded daily, monthly, quarterly, or at other regular intervals. Equation (6.6) is still valid if n is taken to be the number of conversion periods (a *conversion period* being the length of time between interest computations) and i is the interest per conversion period.

EXAMPLE 2.

Find the compound amount on $1,000 invested for four years at 5% compounded semiannually.

The conversion period is six months, or one-half year. The interest rate for one-half year is $i = \frac{5}{2}\%$, or 0.025, and the number of semiannual conversion periods is $n = 8$. The compound amount is

$$C = 1,000(1 + 0.025)^8 = 1,000(1.21840) = 1,218.40.$$

Note the amount here is larger than that of Example 1, where the interest was compounded annually.

The compound amounts at the end of the conversion periods determine a sequence of numbers,

$$P(1 + i), \ P(1 + i)^2, \ P(1 + i)^3, \ P(1 + i)^4, \dots, P(1 + i)^n.$$

Does this look familiar? It is a geometric progression with $r = 1 + i$. These compound amounts form a geometric progression with r greater than one, and therefore increase geometrically. With good reason, Baron Rothschild, who was more than a little familiar with banking, referred to compound interest as the eighth wonder of the world!

The values of $(1 + i)^n$ for various values of i and n are given in Table 6.4 at the end of this chapter. Use of the table eliminates most of the work in determining compound amounts.

The tables at the end of this chapter, although brief, are adequate for our purposes. More complete tables can be found in some of the standard handbooks of mathematical tables.

EXAMPLE 3.

Tom borrowed $2,500 for three years at 6% compounded quarterly. Find the amount he has to pay in three years and the interest that he pays for his loan.

His total payment, including principal and interest, will be the compound amount when $P = 2,500$, $i = \frac{6}{4}$ or $1\frac{1}{2}\%$, and $n = 12$. In Table 6.4, for $i = 1\frac{1}{2}\%$ and $n = 12$, we find the value of $(1 + 0.015)^{12}$ to be 1.19562. Therefore,

$$C = 2,500 \times 1.19562 = 2,989.05.$$

Tom will have to repay $2,989.05. His interest, in dollars, will be

$$2,989.05 - 2,500.00 = 489.05.$$

To determine how much money P should be invested now with compound interest in order to yield a desired amount C, we rewrite Equation (6.6) as

$$\frac{C}{(1 + i)^n} = P,$$

or

$$P = C(1 + i)^{-n}. \tag{6.7}$$

As with simple interest, the quantity P is called the present value of C. Some values of $(1 + i)^{-n}$ are given in Table 6.5 at the end of this chapter.

EXAMPLE 4.

How much money should be invested now at 6% compounded monthly to yield $1,200 two years from now?

$$C = 1,200, i = \tfrac{6}{12} \text{ or } \tfrac{1}{2}\%, n = 24.$$

From Table 6.5, $(1 + 0.005)^{-24} = 0.88719$. Therefore, $P = 1,200 \times 0.88719 = 1,064.63$. $1,064.63 invested now will yield $1,200 in two years.

EXAMPLE 5.

What is the present value at 6% compounded semiannually of $2,400 due the bank in two years?

$$C = 2,400, i = \tfrac{6}{2} \text{ or } 3\%, n = 4.$$

From Table 6.5, $(1 + 0.03)^{-4} = 0.88849$. $P = 2,400 \times 0.88849 = 2,132.38$. The present value is $2,132.38.

EXERCISES

1.† Determine the compound amount of $2,000 invested for four years compounded annually at 6%

   (a) by calculating the interest for each of the four years as in the illustration on page 191.
   (b) by first calculating the value of $(1 + 0.06)^4$ as in Example 1 of this section.
   (c) using Table 6.4.

2. Find the compound amount of

  (a)† $2,500 for four years at 6% compounded semiannually.
  (b) $700 for three years at 8% compounded quarterly.
  (c)† $1,500 for one year at 12% compounded monthly.
  (d) $1,200 for $2\frac{1}{2}$ years at 7% compounded semiannually.

3. Find the total amount to be repaid and the total interest on a loan of

  (a)† $500 for two years at 12% compounded monthly.
  (b) $800 for three years at 6% compounded quarterly.

4.† A father started a savings account for his daughter with a deposit of $200 the day she was born. The account pays 6% compounded quarterly. Determine the amount in the account on her sixth birthday. (Assume that her birthday occurs at the end of a conversion period.)

5. Bob bought a piece of land for $2,750. He expects it to increase in value by about 4% a year. On this basis, what will the land be worth in five years?

6.† What is the present value of $250 at 6% compounded quarterly due in four years?

7. From a financial point of view, which is worth more to you—$600 now that you can invest at 6% compounded semiannually, or $850 in five years?

Table 6.2 gives the account activity for an individual who, for five consecutive years, deposits $1,000 into the account at the end of each year. The money earns 4% interest compounded annually.

TABLE 6.2

End of Year	Deposit	Compound Amount at the End of Five Years
1	1,000	$1,000(1.04)^4 = 1,169.86$
2	1,000	$1,000(1.04)^3 = 1,124.86$
3	1,000	$1,000(1.04)^2 = 1,081.60$
4	1,000	$1,000(1.04)^1 = 1,040.00$
5	1,000	1,000 $= 1,000.00$
		Total: 5,416.32

The last entry in each row gives the compound amount of the deposit in that row. The first deposit, for example, being made at the end of the first year will draw compound interest for four years. The sum of all these compound amounts is the total amount in the account at the end of the fifth year.

Examine carefully the sum that gives this total. In reverse order, it is

$$1,000 + 1,000(1.04) + 1,000(1.04)^2 + 1,000(1.04)^3 + 1,000(1.04)^4.$$

This is the sum of the first five terms of a geometric progression with $a_1 = 1,000$ and $r = 1.04$. Using Equation (6.1), this total should be

$$\frac{1,000[1 - (1.04)^5]}{1 - 1.04},$$

which simplifies to 5,416.32.

Any series of equal payments or deposits made at regular intervals, as in this illustration, is called an *annuity*. This includes car payments, house payments, insurance payments, and so forth. The details of an annuity can be quite varied, because there are different types of annuities. We will consider only the basic type. In our annuities, payments will always be made at the end of the payment period, interest will always be compound interest, and the conversion period will always agree with the payment period—that is, if the payments are monthly, the interest is compounded monthly; if the payments are semiannual, the interest is compounded semiannually, and so on.

The *amount* of an annuity at the end of any payment period is the sum of the compound amounts of the individual payments at that time. The amount of the annuity in the above illustration at the end of five years was $5,416.32. We are interested in a general expression for the amount of an annuity. Suppose that the periodic payments are D dollars each and that the rate of interest per payment period is i. The compound amounts of these payments will form a geometric progression with $a_1 = D$ and $r = (1 + i)$. In the above illustration, D was $1,000 and r was 0.04. The sum of the n terms of the progression will be

$$S = \frac{D[1 - (1 + i)^n]}{1 - (1 + i)} = \frac{D[1 - (1 + i)^n]}{-i},$$

or

$$S = \frac{D[(1 + i)^n - 1]}{i}. \tag{6.8}$$

Equation (6.8) gives the amount of the annuity when the nth payment is made. Use of Table 6.4 to determine the value of $(1 + i)^n$ makes S easy to compute.

EXAMPLE 6.

Find the amount of an annuity of $50 per quarter for three years at 8% compounded quarterly.

The interest per quarter is $i = \frac{8}{4}$ or 2%, $D = 50$, and the number of payments is $n = 12$. The amount, in dollars, is from Equation (6.8):

$$S = \frac{50[(1 + 0.02)^{12} - 1]}{0.02}.$$

From Table 6.4, $(1 + 0.02)^{12} = 1.26824$. Therefore,

$$S = \frac{50[0.26824]}{0.02} = \frac{50[26.824]}{2} = 25[26.824] = 670.60.$$

EXAMPLE 7.

Marilyn, who plans to go to Europe in two years, starts to put $100 per month into a special account which pays 6% compounded monthly. How much will she have for her trip at the end of two years? In this problem, $i = \frac{6}{12}$ or $\frac{1}{2}$%, $D = 100$, and $n = 24$. Therefore,

$$S = \frac{100[(1 + 0.005)^{24} - 1]}{0.005} = \frac{100[0.12716]}{0.005} = \frac{100[127.16]}{5}$$

$$= 20[127.16] = 2{,}543.20.$$

She will have $2,543.20 for her trip.

EXAMPLE 8.

Verify that the college fund illustration at the beginning of this chapter will have approximately $6,180 after 18 years.

The payments constituted an annuity with $D = 200$, $i = 6$%, and $n = 18$. Therefore,

$$S = \frac{200[(1 + 0.06)^{18} - 1]}{0.06}.$$

$(1 + 0.06)^{18} = 2.85434$, so that

$$S = \frac{200[1.85434]}{0.06} = \frac{200[185.434]}{6} = \frac{37{,}086.80}{6} = 6{,}181.13.$$

Although one uses the tables to compute the amount of an annuity to avoid the "dirty" work, one should not lose sight of the fact that we are finding the sum of terms in a geometric progression. In fact, the mathematical model of the compound amounts at the end of the payment periods is the geometric progression. The sum of the terms answers the question, "What is the amount after n periods?" Does this process sound familiar?

EXERCISES

1. Determine the amount of the following annuities:

    (a)† $300 a year for ten years at 5% compounded annually.
    (b) $25 a month for two years at 12% compounded monthly.
    (c)† $60 per quarter for five years at 6% compounded quarterly.

2.† Betty wants to start a small business in five years, at which time she must have $10,000 in capital. At the end of every three months, she puts $400 into an account which pays 6% compounded quarterly. Will she have enough in her account at the end of five years?

3. Payments into a Christmas Club are $15 per month starting December 31 until November 31. On December 5, a check for $180 is sent to the members for their holiday expenses. This is the amount deposited, but includes no interest. How much money would be available for the holidays if the payments would be made into an account which pays 6% compounded monthly?

4.† A manufacturing company expects to replace one of its machines in ten years and estimates that the replacement will cost $250,000. At the end of each year the company puts $18,970 into a special fund to pay for the replacement. If the fund draws 6% interest compounded yearly, verify that at the end of ten years the amount in the fund will pay for the replacement.

6.4 TRUTH IN LENDING

We have seen a variety of ways in which interest can be computed: $1\frac{1}{2}\%$ per month, 6% add-on, 8% discounted, $5\frac{1}{4}\%$ compounded quarterly, and so on. It would be difficult for the ordinary consumer to compare the various methods in order to determine which would be the best for him or her. The Truth-in-Lending law of 1969 remedies this situation. It requires, generally, the disclosure of the sum of all finance charges and the true annual rate of interest. However, the knowledge of the annual rate of interest would be of no advantage if different creditors calculate the annual rate by different methods. The law requires that the declared annual interest rate be determined in the same way by all creditors—and that is the rate at which the loan would be repaid when the payments are first applied to accrued interest and then to reduction of the principal. Recall that this is the manner in which the usual home mortgage loan is handled.

Suppose that a loan of P dollars is to be paid off in n payments of D dollars each. It can be shown, using geometric progressions, that the true annual rate is the value of i which satisfies the following equation:

$$P = \frac{D[1 - (1 + i)^{-n}]}{i}.$$

Tables of values of the expression

$$\frac{1 - (1 + i)^{-n}}{i}$$

have been computed for various values of i and n. To find the correct value of i, one need only consult the tables. However, we shall take an alternate approach that, although it only gives a good approximation to the value of i when the number of payments is not very large, is easy to calculate without the use of tables. The procedure gives a consumer a convenient method of checking whether a declared rate is in fact correct.

Because the interest to be paid at the end of each payment period (usually a month) is determined by the remaining unpaid or outstanding principal during that period, the true rate per payment period would be

$$I = \frac{\text{Interest Paid at the End of the Period}}{\text{Outstanding Principal During the Period}}.$$

This would be true for any payment period; however, the quantities in the division on the right are usually not readily available. Our approximation for the true rate will be obtained by replacing these two quantities by their averages.

During the first payment period—that is, until the first payment is made—the outstanding principal is the total principal P. At the other extreme, the last payment usually is almost all applied toward the principal; there is very little interest to pay. Let us assume that the last payment is all for principal; then the outstanding principal during the last period is D. Recall that D is the amount paid each period. On this basis, the average outstanding principal per payment period is

$$\frac{P + D}{2}.$$

If the loan is to be paid off in n payments, then the total amount paid is the product nD. The difference $nD - P$ gives the total amount of interest paid, so that the average interest per payment period is

$$\frac{nD - P}{n}.$$

Then the interest I, determined by these two averages,

$$I = \frac{nD - P}{n} \div \frac{P + D}{2},$$

gives the approximate interest rate per payment period.

EXAMPLE 1.

A 4.8% add-on loan for $1,000 is to be paid off in a year by equal payments every three months. Use the above procedure to find an approximation to the true annual rate.

The interest will be $48; the payments will be

$$\frac{1,000 + 48}{4} = 262$$

dollars each.

Average interest per payment period:

$$\frac{(4 \times 262) - 1,000}{4} = \frac{1,048 - 1,000}{4} = 12.$$

Average outstanding principal per payment period:

$$\frac{1,000 + 262}{2} = 631.$$

Approximate quarterly interest:

$$I = \frac{12}{631} = 0.019.$$

Approximate annual interest:

$$i = 0.019 \times 4 = 0.076 \text{ or } 7\tfrac{3}{5}\%.$$

To determine if our approximation is reasonable in Example 1, we construct Table 6.3, in which the payments are made in the usual home mortgage loan manner and interest is $7\tfrac{3}{5}\%$. The final payment does not

TABLE 6.3

Payment Number	Outstanding Principal	Interest	Payment on Principal	Remaining Principal Due
1	1,000.00	19.00	243.00	757.00
2	757.00	14.38	247.62	509.38
3	509.38	9.68	252.32	257.06
4	257.06	4.88	257.12	—

agree exactly with the final principal outstanding, but it is not expected to in this case. The rate of 0.076 was supposed to be only a good approximation. The small difference at the end indicates that it is.

EXAMPLE 2.

A $4,200 automobile was purchased with a down payment of $2,000. The balance, including a finance charge of $100, is being paid by monthly payments of $83.00 for three years. Use the above procedure to find an approximation to the true annual rate.

$P = 2,300$, $D = 83$, and $n = 36$.

Average interest per payment period:

$$\frac{(36 \times 83) - 2,300}{36} = \frac{2,988 - 2,300}{36} = 19.11.$$

Average outstanding principal per payment period:

$$\frac{2,300 + 83}{2} = 1,191.50.$$

Approximate monthly interest:

$$I = \frac{19.11}{1,191.50} = 0.016.$$

Approximate annual interest:

$$i = 0.016 \times 12 = 0.192 \text{ or } 19\tfrac{1}{5}\%.$$

EXAMPLE 3.

Jim and Betty obtained a home improvement loan for $1,500 that is to be paid off in monthly payments of $92.52 each for 18 months. At the time they applied for the loan, they were told that the annual rate was 13.5%. Is this the correct annual rate?

$P = 1,500$, $D = 92.52$, $n = 18$.

Average interest per payment period:

$$\frac{(18 \times 92.52) - 1,500}{18} = 9.19.$$

Average outstanding principal per payment period:

$$\frac{1,500 + 92.52}{2} = 796.26.$$

Approximate monthly interest:

$$I = \frac{9.19}{796.26} = 0.0115.$$

Approximate annual interest:

$$i = 0.0115 \times 12 = 0.138 \text{ or } 13.8\%.$$

The declared annual rate of 13.5% is probably correct.

Note that the annual rate required to be specified by the Truth-in-Lending law—in this section called the *true annual rate*—is the rate paid for the money while it is actually used, just as with the usual home mortgage loan. The actual calculation of the amount of interest to be charged can be done in any manner. The law requires only that the corresponding true annual rate be specified.

EXERCISES

Use the procedure discussed in this section in solving the following problems.

1.† Paul is paying off a loan of $1,000 by 18 monthly payments of $60.00 each. Approximate the true annual interest rate that he is paying.

2. If Paul's loan in Exercise 1 were for $900 instead of $1,000, approximate the true annual rate of interest.

3.† Ed, financing a $250 television set through a finance company, is making 15 monthly payments of $19.50 each. Approximate the true annual interest rate that he is paying.

4. Lori bought some furniture for $500, which she is paying for in 12 monthly payments. The amount of the monthly payments was computed based on 7% add-on interest. Approximate the true annual interest rate.

5.† Brad payed $1,500 down on a $3,900 automobile. The balance, along with a finance charge of $150, is being paid off in 30 monthly payments of $106 each. Approximate the true annual interest rate.

6. Suppose that you make a loan for $300, which you are to repay by four payments of $86.54 each every six months. You are told that the true annual interest rate is 12%. Is this the correct rate?

6.5 SUMMARY

In a geometric progression, each term after the first can be obtained by multiplying the preceding term by the common ratio r. If all of the terms are positive and r is greater than 1, the terms in the progression are said to increase geometrically and generally tend to increase very rapidly. If all of the terms are positive with r less than 1, the terms are said to decrease geometrically and generally decrease very rapidly.

The nth term of a geometric progression can be determined by the equation

$$a_n = r^{n-1} a_1$$

and the sum of the first n terms by the equation

$$S_n = \frac{a_1(1 - r^n)}{1 - r} \quad (r \neq 1).$$

Interest is classified into two types: simple and compound. Simple interest, based only on the principal, the rate and the time, is given by

$$I = PRT.$$

The sum of the interest and the principal is called the amount. Using the above equation for interest, an expression for the amount is given by

$$A = P(1 + RT).$$

From this last equation, we get

$$P = \frac{A}{1 + RT},$$

which gives the principal that must be invested at rate R to obtain a given amount A at the end of time T. In this latter case, the principal is called the present value of the amount. R and T must always be expressed in terms of the same unit of time—days, months, years, or so on—in all three of the above equations.

Simple interest is always principal times rate times time, but the manner in which it is calculated, discounted, add-on, home mortgage, or other methods, changes the effective rate—that is, the rate the consumer pays for the money while it is actually used.

Compound interest is simple interest computed again and again. At the end of each conversion period, the current interest is added to the principal. The result is that interest is then paid on interest.

The compound amount, principal plus compound interest (i), at the end of n conversion periods is given by

$$C = P(1 + i)^n.$$

Consequently, the compound amounts at the end of successive conversion periods form a geometric progression with $r = 1 + i$.

From the above equation for compound amount, we get

$$P = C(1 + i)^{-n},$$

which gives the present value of C—that is, the principal P which must be invested for n conversion periods at compound interest i per period to yield the compound amount C. Values of $(1 + i)^n$ and $(1 + i)^{-n}$ are given in Table 6.4 and Table 6.5, respectively, at the end of this chapter.

An annuity is a series of equal payments or deposits made at regular intervals. The amount of a basic annuity at the end of any payment period is the sum of the compound amounts of the individual payments at that time. Because these compound amounts form a geometric progression, the amount of an annuity is a sum of terms of a geometric progression. This sum is

$$S = \frac{D[(1 + i)^n - 1]}{i}.$$

The Truth-in-Lending law requires the disclosure of the true annual rate of interest at the time a loan is made. The true annual rate is the rate at which the loan would be repaid when the payments are first applied to the accrued interest and then to reduction of the principal. This disclosure makes it easier for the consumer to compare the rates charged by various loan institutions.

An easy method of approximating the true rate per payment period is to divide the average interest per payment period

$$\frac{nD - P}{n}$$

by the average outstanding principal per payment period

$$\frac{P + D}{2}.$$

Here, P denotes the principal again, D the amount paid each period, and n the total number of payments.

TABLE 6.4 Values of $(1 + i)^n$

n	$\frac{1}{2}\% = 0.005$	$1\% = 0.01$	$1\frac{1}{2}\% = 0.015$	$2\% = 0.02$	$2\frac{1}{4}\% = 0.025$	n
1	1 00500	1.01000	1.01500	1.02000	1.02500	1
2	1.01003	1.02010	1.03023	1.04040	1.05063	2
3	1.01508	1.03030	1.04568	1.06121	1.07689	3
4	1.02015	1.04060	1.06136	1.08243	1.10381	4
5	1.02525	1.05101	1.07728	1.10408	1.13141	5
6	1.03038	1.06152	1.09344	1.12616	1.15969	6
7	1.03553	1.07214	1.10984	1.14869	1.18869	7
8	1.04071	1.08286	1.12649	1.17166	1.21840	8
9	1.04591	1.09369	1.14339	1.19509	1.24886	9
10	1.05114	1.10462	1.16054	1.21899	1.28008	10
11	1.05640	1.11567	1.17795	1.24337	1.31209	11
12	1.06168	1.12683	1.19562	1.26824	1.34489	12
13	1.06699	1.13809	1.21355	1.29361	1.37851	13
14	1.07232	1.14947	1.23176	1.31948	1.41297	14
15	1.07768	1.16097	1.25023	1.34587	1.44830	15
16	1.08307	1.17258	1.26899	1.37279	1.48451	16
17	1.08849	1.18430	1.28802	1.40024	1.52162	17
18	1.09393	1.19615	1.30734	1.42825	1.55966	18
19	1.09940	1.20811	1.32695	1.45681	1.59865	19
20	1.10490	1.22019	1.34686	1.48595	1.63862	20
21	1.11042	1.23239	1.36706	1.51567	1.67958	21
22	1.11597	1.24472	1.38756	1.54598	1.72157	22
23	1.12155	1.25716	1.40838	1.57690	1.76461	23
24	1.12716	1.26973	1.42950	1.60844	1.80873	24

n	$3\% = 0.03$	$3\frac{1}{2}\% = 0.035$	$4\% = 0.04$	$5\% = 0.05$	$6\% = 0.06$	n
1	1.03000	1.03500	1.04000	1.05000	1.06000	1
2	1.06090	1.07123	1.08160	1.10250	1.12360	2
3	1.09273	1.10872	1.12486	1.15763	1.19102	3
4	1.12551	1.14752	1.16986	1.21551	1.26248	4
5	1.15927	1.18769	1.21665	1.27628	1.33823	5
6	1.19405	1.22926	1.26532	1.34010	1.41852	6
7	1.22987	1.27228	1.31593	1.40710	1.50363	7
8	1.26677	1.31681	1.36857	1.47746	1.59385	8
9	1.30477	1.36290	1.42331	1.55133	1.68948	9
10	1.34392	1.41060	1.48024	1.62889	1.79085	10
11	1.38423	1.45997	1.53945	1.71034	1.89830	11
12	1.42576	1.51107	1.60103	1.79586	2.01220	12

TABLE 6.5 Values of $(1 + i)^{-n}$

i n	$\frac{1}{2}\% = 0.005$	$1\% = 0.01$	$1\frac{1}{2}\% = 0.015$	$2\% = 0.02$	$2\frac{1}{2}\% = 0.025$	i n
1	0.99502	0.99010	0.98522	0.98039	0.97561	1
2	0.99007	0.98030	0.97066	0.96117	0.95181	2
3	0.98515	0.97059	0.95632	0.94232	0.92860	3
4	0.98025	0.96098	0.94218	0.92385	0.90595	4
5	0.97537	0.95147	0.92826	0.90573	0.88385	5
6	0.97052	0.94205	0.91454	0.88797	0.86230	6
7	0.96569	0.93272	0.90103	0.87056	0.84127	7
8	0.96089	0.92348	0.88771	0.85349	0.82075	8
9	0.95610	0.91434	0.87459	0.83676	0.80073	9
10	0.95135	0.90529	0.86167	0.82035	0.78120	10
11	0.94661	0.89632	0.84893	0.80426	0.76214	11
12	0.94191	0.88745	0.83639	0.78849	0.74356	12
13	0.93722	0.87866	0.82403	0.77303	0.72542	13
14	0.93256	0.86996	0.81185	0.75788	0.70773	14
15	0.92792	0.86135	0.79985	0.74301	0.69047	15
16	0.92330	0.85282	0.78803	0.72845	0.67362	16
17	0.91871	0.84438	0.77639	0.71416	0.65720	17
18	0.91414	0.83602	0.76491	0.70016	0.64117	18
19	0.90959	0.82774	0.75361	0.68643	0.62553	19
20	0.90506	0.81954	0.74247	0.67297	0.61027	20
21	0.90056	0.81143	0.73150	0.65978	0.59539	21
22	0.89608	0.80340	0.72069	0.64684	0.58086	22
23	0.89162	0.79544	0.71004	0.63416	0.56670	23
24	0.88719	0.78757	0.69954	0.62172	0.55288	24

i n	$3\% = 0.03$	$3\frac{1}{2}\% = 0.035$	$4\% = 0.04$	$5\% = 0.05$	$6\% = 0.06$	i n
1	0.97087	0.96618	0.96154	0.95238	0.94340	1
2	0.94260	0.93351	0.92456	0.90703	0.88999	2
3	0.91514	0.90194	0.88899	0.86384	0.83962	3
4	0.88849	0.87144	0.85480	0.82270	0.79209	4
5	0.86261	0.84197	0.82193	0.78353	0.74726	5
6	0.83748	0.81350	0.79031	0.74622	0.70496	6
7	0.81309	0.78599	0.75992	0.71068	0.66506	7
8	0.78941	0.75941	0.73069	0.67684	0.62741	8
9	0.76642	0.73373	0.70259	0.64461	0.59190	9
10	0.74409	0.70892	0.67556	0.61391	0.55839	10
11	0.72242	0.68495	0.64958	0.58468	0.52679	11
12	0.70138	0.66178	0.62460	0.55684	0.49697	12

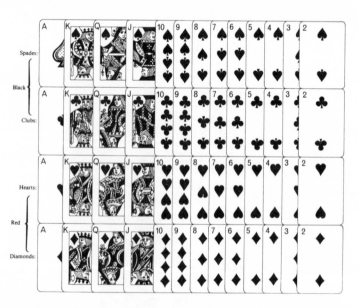

An ordinary deck of 52 playing cards is the basis for a number of
examples and exercises in this chapter.

chapter 7

probability II

This chapter expands upon the introduction to probability given in Chapter 2. Our major concern will be methods for calculating probabilities without actually constructing the sample spaces. We shall also consider additional methods for determining probability functions.

7.1 PROBABILITY FUNCTIONS

Recall that each application of the system of probability involves two basic items, a sample space and a probability function. Probability functions arise in a number of ways; those considered here are of three types— probability functions for equilikely outcomes, probability functions determined by past observations, and subjective probability functions.

Probability functions for equilikely outcomes are applicable for experiments in which any one element of the sample space is as likely to occur as any other. (Such an experiment is the roll of a single balanced die.) *If there are n events in the sample space for the experiment, each element in the sample space is assigned a probability of 1/n.* In this way, each element in the sample space is given the same probability in such a

way that the sum of the probabilities assigned is 1. In the single die situation, because there are 6 elements in the sample space, each element is
assigned the probability of $\frac{1}{6}$. Most of the probability functions of Chapter 2 were of this type.

EXAMPLE 1.

A card is drawn at random from an ordinary deck of 52 playing cards.
What is the probability that it is a picture card (jack, queen, king,
or ace)?

The sample space consists of a listing of the 52 cards. Because the
drawing is random, the 52 elements are equilikely and therefore each
is given a probability of $\frac{1}{52}$. The event of getting a picture card contains 16 elements—4 jacks, 4 queens, 4 kings, and 4 aces. The prob-
ability of getting a picture card is therefore $16(\frac{1}{52}) = \frac{16}{52} = \frac{4}{13}$.

If the elements of the sample space are not equilikely, the method of
determining probabilities just described would not be realistic. One method
of determining the probability function is to study outcomes of past occurrences of the experiment.

Suppose an experiment consists of a student taking a particular mathematics course, and the interest is in whether the student, randomly chosen,
will pass or fail the course. The sample space would be $\{P, F\}$. What probabilities should be assigned to P and to F?

Suppose also that 500 students took the same course from the same
instructor during the past year, and that, of the 500, 450 passed and 50
failed the course. On this basis, a probability of $\frac{450}{500} = \frac{9}{10}$ would be assigned
P and $\frac{50}{500} = \frac{1}{10}$ would be assigned F. Note that *each element in the sample
space is assigned a probability equal to the proportion of times that ele-
ment occurred during the past observations.*

EXAMPLE 2.

In determining whether a particular bus arrives at a certain location
on time, the following sample space could be used:

$$S = \{(-\infty, -5), [-5, -1), [-1, 0), [0], (0, 1], (1, 5], (5, \infty)\}$$

where the elements describe the arrival time of the bus as follows.

$(-\infty, -5)$ — Late, more than five minutes.
$[-5, -1)$ — Late, more than one minute and up to five minutes,
  including five minutes.
$[-1, 0)$ — Late, up to one minute, including one minute.
$[0]$ — On time.
$(0, 1]$ — Early, up to one minute, including one minute.
$(1, 5]$ — Early, more than one minute and up to five minutes,
  including five minutes.
$(5, \infty)$ — Early, more than five minutes.

It would not be very reasonable to expect that each element of the sample space would be as likely to occur as any other. Suppose, instead, that the bus is observed for a period of 60 days and its arrival times were tabulated as follows:

Element	Number of Occurrences
$(5, \infty)$	4
$(1, 5]$	6
$(0, 1]$	15
$[0]$	6
$[-1, 0)$	10
$[-5, -1)$	15
$(-\infty, -5)$	4

On this basis, the respective probabilities would be $\frac{4}{60} = \frac{1}{15}, \frac{1}{10}, \frac{1}{4}, \frac{1}{10}, \frac{1}{6}, \frac{1}{4}, \frac{1}{15}$.

What is the probability that tomorrow the bus is early?

The event "the bus is early" contains the elements $(0, 1], (1, 5]$, and $(5, \infty)$; therefore, the probability of early arrival would be $\frac{1}{4} + \frac{1}{10} + \frac{1}{15} = \frac{5}{12}$.

The danger in this type of probability function lies in the question of the extent to which past behavior indicates future behavior! If the supervisor of the bus driver in Example 2 began to insist that the driver be more prompt, then the probability of $\frac{5}{12}$ that the bus is early may no longer be valid. Or, if one were to assign probabilities to the various teams of the National Football League for winning the next league championship based on last season's performance, the results might not be very realistic if there have been significant changes in the coaching staffs and/or player rosters.

Finally, if the equilikely method is not applicable and there is no past experience to rely upon, a remaining possibility is to assign probabilities to the elements of the sample space on the basis of the *subjective opinion* of someone familiar with the details of the experiment—essentially an educated guess.

Suppose five horses run in a race; the sample space for the winning of the race would be {1, 2, 3, 4, 5}. It is not likely that each horse has the same chance of winning. Supposing also that none of these five horses ever appeared together in a previous race; there is no past experience to rely upon. Someone who is familiar with the five horses would be in a position to give an opinion as to what probabilities to assign to each horse's winning. Probability functions determined in this manner fall into the class of subjective probability functions.

Or, suppose a company is about to market a new product and that it hopes to sell 150,000 units during the first year. The company can make a profit from the product if it charges at least $2.00 per unit, but wants to make a larger profit if it can still accomplish its sales objective of 150,000 units. Company executives who are familiar with the market behavior are asked to give their estimate of the probability of attaining the sales objective if the price is set at $2.00, $2.25, and $2.45. The sample space is then {2.00, 2.25, 2.45}. Probabilities are assigned on the basis of the subjective opinions of the executives, and the final decision is made accordingly.

EXAMPLE 3.

Mark, a graduating senior, is interested in the probability of whether he can get into law school. His advisor estimates the probability to be 0.65 that he will be admitted. Based on this subjective probability, what is the probability that he will not be admitted?

The probability that he will not be admitted is $1 - 0.65$, or 0.35.

Care should be taken in the choice of the probability function, as the probability of each event in the sample space is determined by the manner in which the probability function assigns values to each of the elements in the sample space. Therefore, these probabilities are realistic only to the extent that the probability function is realistic!

EXAMPLE 4.

Each ticket in a state lottery contains two four-digit numbers (from 0000 to 9999). The drawing consists of randomly choosing one number between 0000 and 9999. Each person who holds a ticket with one of the numbers being the same as that drawn is a winner.

Susan has a ticket with the numbers 0531 and 2507. What is the probability that she will be a winner?

The sample space has 10,000 elements—the numbers from 0000 to 9999. How should probabilities be assigned to these elements? Are they equilikely so that each element should be given the same probability? Are they not equilikely so that probabilities should be assigned based on past drawings? Or, should we attempt to find an educated guess?

As long as the lottery is honest, the first type of probabilities should be assigned; each element is as likely to occur as any other. Two of these, 0531 and 2507, are in the event "Susan wins." The corresponding probability is therefore $2 \times \frac{1}{10,000} = \frac{1}{5,000}$.

EXERCISES

1.[†] What is the probability that the bus in Example 2 of this section will not be late?

2. In the state lottery of Example 4 of this section, Gene has a ticket with the numbers 0483 and 5361.
    (a) What is the probability that either Susan or Gene will win at the drawing?
    (b) What is the probability that both Susan and Gene will win at the drawing?

3.[†] What is the probability that an individual in Toledo will receive less than three telephone calls per day, if a survey of the telephone subscribers in Toledo for six months indicates that the probabilities of an individual receiving 0, 1, 2, 3, 4, or 5 or more phone calls per day are, respectively, 0.10, 0.25, 0.25, 0.15, 0.15, 0.10?

4. The letters of the word *mathematics* are written on eleven cards, one letter per card. The cards are placed in a hat and one is chosen at random. What is the probability that the card drawn contains a letter that precedes N in the alphabet?

5.[†] A card is drawn at random from an ordinary deck of 52 playing cards. What is the probability that the card drawn is a 5?

6. Paul has gotten into the habit of counting the number of cars on the commuter train which he rides to work each day. During eight weeks (40 work days), there were 12 cars on 14 days, 13 cars on 12 days, 14 cars on 10 days and 15 cars on 4 days. On this basis, what is the probability that the train will have more than 12 cars next Monday?

7.[†] Stretch Rubber Company tested 1,000 tires of a particular type. Of these, 750 maintained a certain depth of tread for more than 40,000 miles and 250 did not. What is the probability that one of this type of tire bought by a customer will maintain the same depth of tread?

8. You have two pairs of shoes in a dark closet. You randomly pick two of the shoes. What is the probability that you have a pair?

7.2 COUNTING TECHNIQUES AND PROBABILITY

Often the elements in a sample space, although finite in number, are too numerous to list individually. For instance, there are over 2 million possible five-card hands that can be dealt from an ordinary deck of 52 playing cards.

It would be virtually impossible to manually list all of the elements of the sample space for this experiment.

However, when the elements are equilikely, it is sufficient to know only how many elements are in a sample space and in the events in which we are interested. There is no need to actually construct the sample space. (You may have noticed that we omitted the sample space in some of the examples in the previous section.) The counting techniques of this section are used to determine the number of elements in sample spaces and events without constructing them.

There are two basic techniques to be considered: the *fundamental counting principle* and *combinations*.

Suppose a salesman's route requires that he travel from Toledo to Detroit and then to Chicago. Suppose, further, that he can choose one of two highways (*A* or *B*) from Toledo to Detroit, and one of three highways (*R*1, *R*2, or *R*3) from Detroit to Chicago. In how many ways can he pick his route?

We illustrate the possible routes by means of a *tree graph* (Fig. 7.1):

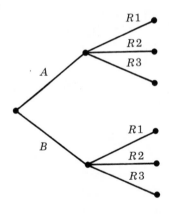

FIGURE 7.1

If he picks route *A* to get to Detroit, he has a choice of three highways to get to Chicago. Similarly, if he chooses route *B*. His possible routes correspond to the paths through the tree starting at the left node and ending at one of the nodes on the right. His first choice can be made in two ways; his second in three. Altogether, he has 2 × 3 = 6 different possible choices.

This is a special case of the fundamental counting principle, which states:

If one thing can happen in n_1 ways and, for each of these n_1 ways, a second thing can happen in n_2 ways, the two things can occur in order in $n_1 \times n_2$ ways.

EXAMPLE 1.

A student must take one of three mathematics courses at 9:00 and one of four English courses at 10:00. In how many ways can he arrange his 9:00–10:00 schedule?

Because the first thing, choosing a mathematics course, can happen in three ways and the second, choosing an English course, can happen in four ways, together they can happen in $3 \times 4 = 12$ ways.

The fundamental counting principle can be extended to counting the number of ways more than two things can happen. Suppose the salesman referred to previously could return from Chicago to Toledo by two different highways, and we are to determine the number of routes at his disposal for the round trip (Fig. 7.2). The final 12 branches correspond to the number of paths through the tree, or the number of possible routes that he can choose.

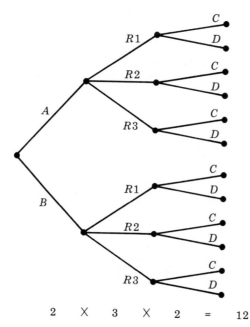

$$2 \quad \times \quad 3 \quad \times \quad 2 \quad = \quad 12$$

FIGURE 7.2

To determine the number of ways that three events can occur in order, the fundamental principle states:

If one thing can happen in n_1 ways, if for each of these n_1 ways a second thing can then happen in n_2 ways, and if for each of the $n_1 \times n_2$ ways in which these two things can happen in order, a third thing can then happen in n_3 ways, the three things can happen in sequence in $n_1 \times n_2 \times n_3$ ways.

EXAMPLE 2.

How many three-digit numbers can be formed using the ten digits 0, 1, 2, 3, 4, 5, 6, 7, 8, 9?

The construction of a three-digit number involves three things—the choosing of the first digit, which can happen in any one of ten ways; the choosing of the second digit, which can also happen in one of ten ways; and the choosing of the third digit, which can also happen in one of ten ways. Hence, the total number of such digits is, not unexpectedly, $10 \times 10 \times 10 = 1,000$.

EXAMPLE 3.

Suppose that no digit can be used twice in the construction of the above three-digit numbers. Then how many such numbers can be formed?

The first choice can still be made in one of ten ways, but after the choice of the first digit there are only nine ways to choose the second digit, and then eight ways to choose the third digit. Hence, there are $10 \times 9 \times 8 = 720$ three-digit numbers without repetition of any one digit.

EXAMPLE 4.

A particular type of combination lock has twelve numbers on it. How many combinations of four different numbers can be formed to open the lock?

The first number can be chosen in twelve ways, the second in eleven, the third in ten and the fourth in nine. Therefore, there are $12 \times 11 \times 10 \times 9 = 11,880$ different combinations.

EXAMPLE 5.

A mathematics honorary fraternity consists of 15 students, 8 girls and 7 boys. In how many ways can the fraternity choose a president, vice-president, treasurer, and secretary?

The president can be chosen in fifteen ways, the vice-president in fourteen ways, and so forth, so that the officers can be chosen in a total of

$$15 \times 14 \times 13 \times 12 = 32,760$$

different ways.

EXAMPLE 6.

In how many ways can the fraternity of Example 5 above pick their officers if they agree that the president and secretary must be a girl and that the vice-president and treasurer must be a boy?

In that case, there would be

$$8 \times 7 \times 7 \times 6 = 2{,}352$$

different slates of officers.

Recall that we are interested in the counting techniques from the point of view of calculating probabilities. The next two examples illustrate the use of the fundamental principle in probability.

EXAMPLE 7.

If a three-digit number is chosen at random from the numbers 000 to 999, what is the probability that the number contains three different digits?

The sample space consists of 1,000 equilikely elements—the numbers from 000 to 999, each with a probability of $\frac{1}{1,000}$. From Example 3, the event of interest has 720 elements; its probability is therefore $720 \times \frac{1}{1,000}$.

EXAMPLE 8.

Kay, Betty, and Joanne are three of the girls in the fraternity of Example 6 above. If the choice of officers is made by lot, with the distinction by sex as indicated, what is the probability that two of these girls will be officers?

The sample space would consist of the 2,352 different slates of officers determined in Example 6.

By the fundamental counting principle, the event that Kay, Betty, and Joanne fill the president and secretary positions contains

$$3 \times 2 \times 7 \times 6 = 252$$

elements. The probability of this event is, therefore,

$$252 \times \frac{1}{2,352} = \frac{3}{28}.$$

Before proceeding to the exercises, note that the fundamental counting principle is concerned with *order*—counting the number of ways things can happen in order. This includes the number of ordered arrangements which can be made from a specified number of objects chosen from some given set. The four numbers to be chosen from twelve in Example 4 were to be in a specific order. Any attempt to open a combination lock with the correct numbers but in wrong order is doomed to failure!

Also, the four officers to be chosen from fifteen in Example 5 were to be in a specific order—president, vice-president, secretary, and treasurer. If the four persons chosen interchanged offices, the result would be a different slate of officers.

It will be important to appreciate this order concept in the remainder of this section.

EXERCISES

1. Every television set manufactured by a particular company is classified by picture, style, and quality in the following manner:

   by picture — black-white or color;
   by style   — floor, table, or portable; and
   by quality — good, medium, or deluxe.

   Draw a tree graph to illustrate the 18 different models of television sets manufactured.

2.† Susan has four sweaters and six skirts. How many different sweater-skirt outfits can she select to wear?

3. A mathematics club has twelve members. In how many ways can the club select a president and vice-president?

4.† A gourmet club has twelve members, seven women and five men. In how many ways can the club select a president and vice-president if the president must be a woman and the vice-president must be a man?

5. The Greek alphabet consists of twenty-four letters. How many names of fraternities can be formed if each name consists of three letters and

   (a) no repetitions of letters are allowed?
   (b) repetitions of letters are allowed?

6.† How many batting orders are possible for players on a baseball team if

   (a) there is no designated pinch hitter and the pitcher must bat last?
   (b) there is a designated pinch hitter?

   (There are nine players on a baseball team. In addition, a tenth player may be designated as a pinch hitter, who then bats for the pitcher and who may occupy any position in the batting order.)

7. How many four-digit integers less than 5,000 can be formed using the digits 2, 3, 4, 5, and 6 if

   (a) repetitions of digits are allowed?
   (b) repetitions of digits are not allowed?

8.[†] In a particular state, each license plate contains two letters followed by four digits. If no digit can be used twice in any one license plate, how many different license plates can be made?

9. Six books, two of which are mathematics books, are to be placed on a shelf. If the two mathematics books are to be placed together, in how many ways can the books be arranged

   (a) if the two mathematics books must be to the left of the others?
   (b) without the restriction in part (a)? (*Hint:* First determine the number of positions which the two mathematics books can occupy; then consider the arrangement to be a three-step process.)

10.[†] An experiment consists in rolling four dice (white, red, green, and yellow).

   (a) How many elements are there in the sample space? Assume the elements are of the form (2431), which would indicate that a 2 is obtained on the white, a 4 on the red, a 3 on the green, and a 1 on the yellow.
   (b) How many elements in the above sample space are in the event "a 2 is obtained on the white die?"
   (c) What is the probability of obtaining a 2 on the white die?
   (d) How many elements in the above sample space are in the event "a 2 is obtained on the white die and a 1 on the yellow?"
   (e) What is the probability of obtaining a 2 on the white die and a 1 on the yellow?

11. The four aces, four kings, and four queens are removed from an ordinary deck of playing cards and shuffled. (The remaining cards are discarded.) An experiment consists of drawing four cards, with each card being replaced after it is drawn.

   (a) How many elements are there in the sample space?
   (b) How many elements are there in the event "the first three cards drawn are aces but the fourth is not an ace?"
   (c) What is the probability that the first three cards drawn (but not the fourth) are aces?
   (d) How many elements are there in the event "the first three or the first four cards drawn are aces?"
   (e) What is the probability that the first three or first four cards drawn are aces?

*12. Use the fundamental principle to verify that there are 2^n subsets of a set which contains n elements. (*Hint:* Consider a subset to be determined by making a "yes" or "no" decision about each of the elements.)

If an organization has twelve members and wants to send a committee of two to the national convention, how many different committees can be formed? Committees do not have structure, so we are no longer interested in ordered arrangements—the committee made up of Mary and Larry is the same as the committee made up of Larry and Mary. Consequently, we now want to know how many subsets of two members can be formed without regard to order. Such unordered subsets are called *combinations*.

Our next objective is to establish a method for counting the number of possible combinations. To do so, we shall have to consider counting ordered arrangements from another point of view.

In Example 3, we were interested in how many three-digit numbers could be constructed using the digits 0 through 9 without repetition. There were $10 \times 9 \times 8 = 720$ different such numbers. (Ordered arrangements of this type without repetitions are sometimes called *permutations*.)

Because

$$10 \times 9 \times 8 = 10 \times 9 \times 8 \times \frac{7 \times 6 \times 5 \times 4 \times 3 \times 2 \times 1}{7 \times 6 \times 5 \times 4 \times 3 \times 2 \times 1}, \qquad (7.1)$$

the right side of this last equation could be used for the calculation as well as the left. The merit of this second form is that the notation can be simplified by use of *factorials*.

The product $n \times (n-1) \times (n-2) \times \cdots \times 3 \times 2 \times 1$, for any natural number n, is written $n!$ [read "n factorial"]. It follows, for example, that

$5! = 5 \times 4 \times 3 \times 2 \times 1 = 120$
$4! = 4 \times 3 \times 2 \times 1 = 24$
$7! = 7 \times 6 \times 5 \times 4 \times 3 \times 2 \times 1 = 5,040$

Now we can write the right-hand side of Equation (7.1) as

$$\frac{10!}{7!}.$$

Note that the numerator corresponds to the number of objects available, and the denominator corresponds to the number not used.

In general, the number of ordered arrangements without repetitions that can be made using r objects when there are n to choose from is given by

$$\frac{n!}{(n-r)!}.$$

In the above illustration, n is 10 and r is 3 so that $n - r$ is 7. It can happen that all n objects are to be used, in which case $r = n$. The denominator in the above expression then becomes $0!$. Because 0 is not a natural number, $0!$ has no meaning from the definition just given for factorials.

By the fundamental counting principle, the number of ordered arrangement of n objects when all are used is $n!$. If $0!$ has the value 1, then

$$\frac{n!}{0!} = n!.$$

Therefore, we define $0!$ to be equal to 1, and the expression

$$\frac{n!}{(n-r)!} \qquad (7.2)$$

can be used to count the number of ordered arrangements without repetitions whether or not n equals r. For example, if $n = 7$ and $r = 4$, then Equation (7.2) becomes

$$\frac{7!}{(7-4)!} = \frac{7!}{3!} = \frac{7 \times 6 \times 5 \times 4 \times 3 \times 2 \times 1}{3 \times 2 \times 1} = 840.$$

Or, if $n = 4$ and $r = 4$, we get

$$\frac{4!}{(4-4)!} = \frac{4!}{0!} = \frac{4 \times 3 \times 2 \times 1}{1} = 24.$$

We return now to the method of counting combinations. We can expect there to be fewer combinations than ordered arrangements. This is because there are usually many ordered arrangements corresponding to one subset. For a specific example, consider how many ordered arrangements of the numbers in the set {5, 7, 9} are possible. By the fundamental counting principle, there should be $3 \times 2 \times 1 = 6$; these six are

579 759 957
597 795 975.

To the one set of 3 elements there are $3!$ corresponding ordered arrangements. In the same way, to any one set of r elements, there correspond $r!$ ordered arrangements. Therefore, to determine the number of subsets of r objects (without regard to order) which can be chosen from a given set of n objects, it is sufficient to divide the number of ordered arrangements of n objects chosen r at a time by $r!$; that is, the number of such subsets of r objects is

$$\frac{n!}{(n-r)!} \div r!$$

or

$$\frac{n!}{r!(n-r)!}$$

The notation used for this last expression is $C(n,r)$ [read "the number of combinations of n objects chosen r at a time"]. That is,

$$C(n,r) = \frac{n!}{r!(n-r)!}.$$

EXAMPLE 9.

Returning to the opening illustration of this discussion, how many committees of two can be formed from the membership of an organization if there are a total of twelve members?

In this case, $n = 12$ and $r = 2$. There are

$$C(12,2) = \frac{12!}{2!\,10!} = \frac{12 \times 11 \times 10!}{2 \times 1 \times 10!} = 66$$

possible committees.

EXAMPLE 10.

How many different five-card hands can be dealt from an ordinary deck of 52 playing cards?

Because we are not concerned with the order in which the cards are dealt, the number of such possible hands is

$$C(52,5) = \frac{52!}{5!\,47!} = 2{,}598{,}960.$$

Why should we not use the fundamental counting principle in Example 10? If we did, with the first thing happening 52 ways, the second 51, and so forth, to obtain

$$52 \times 51 \times 50 \times 49 \times 48 = 311{,}875{,}200,$$

we would have the number of ways the five cards could be dealt in order or could be held in your hand in order. Normally, the important thing in playing cards is just the five cards you have; there is no concern about order. Consequently, one would be interested only in the number of sets of five cards.

EXAMPLE 11.

A bag contains 14 balls, 8 red and 6 white. Suppose an experiment consists of drawing out three of the balls. How many elements are there in the sample space?

As long as the order in which the balls are drawn is of no concern, the possible outcomes of the experiment would be all combinations of three balls. The number of such combinations is

$$C(14,3) = \frac{14!}{3!\,11!} = 364.$$

Note that in Example 11, we did not distinguish between the red and the white; all 14 balls were treated on an equal basis. If we were interested in obtaining the number of ways of obtaining two red and one white, then we would have had to count within the red balls and within the white balls separately. In general, if counting combinations involves more than one group of objects, all of which are not treated on exactly the same basis, then combinations must be first counted within each group and then the results combined by the fundamental principle.

EXAMPLE 12.

In the sample space of Example 11, how many elements are there in the event "two red balls and one white ball are drawn"?

The number of ways that two red balls can be drawn is $C(8,2) = 28$; the number of ways that one white ball can be drawn is $C(6,1) = 6$. Together, these two outcomes can occur in $28 \times 6 = 168$ ways.

EXAMPLE 13.

What is the probability of drawing two red balls and one white ball in Example 11?

There are 364 elements in the sample space (Example 11). Of these, 168 are in the event of two red and one white (Example 12). The probability is therefore $\frac{168}{364} = \frac{6}{13}$.

EXAMPLE 14.

What is the probability of obtaining three aces and two queens in a five-card hand dealt from an ordinary deck of playing cards?

The sample space consists of the 2,598,960 possible five-card hands determined in Example 10. For the event of interest, the three aces can be picked in $C(4,3) = 4$ ways and the two queens in $C(4,2) = 6$ ways. Therefore, the event contains $4 \times 6 = 24$ elements. Its probability is

$$24 \times \frac{1}{2,598,960},$$

or approximately 0.000009 (pretty small!).

In concluding this section, recall again that the purpose of the counting procedures is to determine probabilities without actually constructing the sample space. This applies only when all of the elements in the sample space are equilikely. All one needs to know is the number of elements in the sample space and in the event of interest.

It is not always immediately obvious which procedure to use in a particular problem. This is due to the fact that there are no general guidelines on how to handle every possible case. Like the word problems in algebra, each case must be analyzed separately. However, if counting techniques are to be used, usually the first question to be answered is whether ordered arrangements or combinations are called for—and if combinations, whether the objects must be treated as separate groups. Normally, these questions must be answered from the nature of the situation being examined, and not from the wording of the problem or exercise. Moreover, the meaning of the word *ordered* varies from one situation to the next.

EXERCISES

In Exercises 2 through 13, use the method of combinations and/or the fundamental counting principle as appropriate.

1.† Evaluate: $C(12,3)$, $C(7,7)$, $C(9,1)$.

2. An automobile dealer receives twelve new cars. He wants to display three in his showroom. How many groups of 3 can he choose to show?

3.† Peano's Pizza Parlor makes sausage, anchovie, mushroom, pepper, onion, pepperoni, and hamburger pizzas. Peano's Columbus Day special consists of combinations of any three toppings for the price of one. In how many ways can the Columbus Day special be made?

4. A quiz consists of five questions, of which the student is to answer three. In how many ways can the student pick the three questions to be answered?

5.† Ten men on a football team are all able to play each of the four backfield positions. How many different backfield line-ups are available at any one time?

6. In the United States Congress, a conference committee is to be composed of three senators and three representatives. In how many ways can this be done? (There are 435 representatives and 100 senators.)

7.† A class consists of seven boys and seven girls. If the boys must sit in the odd-numbered seats and there are 14 seats in the room, how many different seating arrangements are possible?

8. There are fourteen cars in a parking lot, nine domestic and five foreign. An experiment consists in picking out three of the cars at random.
    (a) How many elements are there in the sample space?
    (b) How many elements are there in the event "all three cars chosen are domestic"?
    (c) What is the probability that all three chosen are domestic?

9.† (a) A department store manager has seven qualified applicants available for two saleswomen positions in a certain department and five qualified applicants for three salesmen positions in the same department. In how many ways can he fill the five positions?

(b) If Bob and Ted are two of the qualified applicants, what is the probability that they will both be hired, assuming that the possible combinations of five applicants are equilikely?

10. (a) A television executive must schedule three half-hour programs in three open time slots of a particular evening's program. If he has seven such half-hour programs to choose from, in how many ways can he design the evening's program?

   (b) Suppose that the seven available programs are all of about the same quality, so that the choice of programs is essentially a random choice. If one of the seven is a quiz program, what is the probability that the quiz program is chosen to be shown? (*Hint:* Consider first the number of time slots in which the quiz program can be shown.)

11.† What is the probability of obtaining exactly three queens in a five-card hand dealt from an ordinary deck of playing cards?

12. Three balls are drawn from a bag containing four red, four blue, and four green balls. What is the probability of obtaining two blue and one green?

*13.† What is the probability of obtaining "three of a kind" (three matching cards with the other two cards neither matching those three nor each other) in a five-card hand dealt from an ordinary deck of playing cards?

14. We noted in passing that ordered arrangements that are all taken from the same set are called *permutations* if repetitions are not allowed. The notation for the number of such arrangements of n objects using r at a time is $P(n,r)$. Equation (7.2) gives the value of $P(n,r)$:

$$P(n,r) = \frac{n!}{(n-r)!}$$

   (a) Rework Exercise 3 of the previous set of exercises by counting permutations.

   (b) Rework Exercise 5(a) of the previous set of exercises by counting permutations.

   (c) Are combination locks misnamed? (See Example 4 of this section.)

7.3 THE MULTIPLICATION RULE

Up to this point, we have added probabilities in a variety of situations, but we have never *multiplied* them. Multiplication of probabilities enters into determining the probability of each of several events occurring. To see how this is done, we shall need the concept of *conditional probability*.

To illustrate this concept, suppose that a car is chosen at random from the lot of an auto sales company that handles both foreign and domestic cars. Let F be the event "a foreign car is chosen" and E be the event "a sports car is chosen." The probability that E occurs, given that F occurs—that is, the probability that a sports car is chosen given that a foreign car is chosen—is an example of a conditional probability. The notation used to indicate this type of probability is $P(E|F)$ [read "the probability of E given F"].

To calculate this conditional probability, suppose that the company has on its lot 36 cars classified as follows:

	Domestic	Foreign
Sedans	18	6
Sports Cars	4	8

The corresponding sample space would have 36 elements representing the 36 cars. However, the condition that F occurs reduces the choice to the foreign cars.

	Foreign
Sedans	6
Sports Cars	8

The corresponding reduced sample space has just 14 elements, and the event E of choosing a sports car in this reduced sample space has 8 elements. Therefore,

$$P(E|F) = 8 \times \tfrac{1}{14} = \tfrac{8}{14}, \text{ or } \tfrac{4}{7}.$$

Examining the table for all 36 cars, we see that 8 is the number of elements in the event $E \cap F$ (the car chosen is both a sports car and a foreign car). Also, 14 is the number of elements in the event F (a foreign car is chosen). Consequently, we might expect that the conditional probability

$$P(E|F) = \tfrac{8}{14}$$

could just as well have been determined in the sample space for all 36 cars. In fact, $P(E \cap F) = \tfrac{8}{36}$ and $P(F) = \tfrac{14}{36}$, so that

$$\frac{P(E \cap F)}{P(F)} = \frac{8}{36} \div \frac{14}{36} = \frac{8}{36} \times \frac{36}{14} = \frac{8}{14};$$

that is,

$$\frac{P(E \cap F)}{P(F)} = P(E|F).$$

By multiplying each side of this last equation by $P(F)$, we get the more useful form

$$P(E \cap F) = P(F) \cdot P(E|F).$$

This equation is referred to as the *Multiplication Rule*. It gives a method for determining the probability that *each of two events*, E and F, occurs.

Use of the Multiplication Rule is not the only way of determining the probability that each of two events occurs. However, it often simplifies the sample spaces and/or calculations required.

EXAMPLE 1.

Two cards are drawn at random from an ordinary deck of playing cards. The first is not replaced before the second is drawn. What is the probability that both are aces?

Let F be the event "an ace is drawn on the first draw" and E be the event "an ace is drawn on the second draw." We wish to determine $P(E \cap F)$. The sample space for the first draw contains 52 elements, 4 of which are in the event F; hence $P(F) = \frac{4}{52}$, or $\frac{1}{13}$. The sample space for the second draw contains 51 elements, 3 of which are in the event E, assuming that an ace was drawn on the first try; hence, $P(E|F)$, the probability of an ace on the second draw given that an ace is drawn on the first, is $\frac{3}{51}$, or $\frac{1}{17}$. By the Multiplication Rule

$$P(E \cap F) = \frac{1}{13} \times \frac{1}{17} = \frac{1}{221}.$$

In Example 1, note that the sample space relative to the outcome of event F contains only 52 elements and that of $E|F$ contains 51. This leads to relatively simple calculations as compared to those in the sample space of all possible two-card combinations which consists of $C(52,2) = 1,326$ elements. Of these 1,326, $C(4,2) = 6$ consist of two aces. The probability of the two-ace draw in this sample space is $6 \times \frac{1}{1,326}$, which again equals $\frac{1}{221}$.

EXAMPLE 2.

If the first card in Example 1 is replaced before the second is drawn, what is the probability of drawing two aces?

Again, $P(F) = \frac{1}{13}$. If the first card is replaced before the second is drawn, there are again four aces in the 52 cards so that $P(E|F)$ is now $\frac{1}{13}$. By the Multiplication Rule

$$P(E \cap F) = \frac{1}{13} \times \frac{1}{13} = \frac{1}{169}.$$

EXAMPLE 3.

A coin and die are tossed simultaneously. What is the probability of obtaining a Head on the coin and a 5 on the die?

We are interested in each of the two events:

  E: a Head is obtained on the coin, and
  F: a 5 is obtained on the die

occurring. Because $P(F) = \frac{1}{6}$ and $P(E|F) = \frac{1}{2}$, $P(E \cap F) = P(F) \cdot P(E|F)$ $= \frac{1}{6} \times \frac{1}{2} = \frac{1}{12}$.

In the last example, note that $P(E|F) = P(E)$; this reflects the fact that the outcome of the die does not affect the outcome of the coin. Such events are called *independent* events; otherwise they are called *dependent*. Example 1 above was concerned with dependent events, Example 2 with independent.

In the case of independent events, the Multiplication Rule takes the simple form.

  $P(E \cap F) = P(F) \cdot P(E)$;

that is, the probability of the occurrence of each of two independent events is the product of their individual probabilities.

EXAMPLE 4.

Two friends, Anne and Laura, are each expecting their first baby. What is the probability that they each have girls, assuming that the probability of a female baby in any instance is $\frac{1}{2}$?

If F is the event that Anne has a girl and E the event that Laura has a girl, then E and F are independent events. The probability that both occur is therefore

  $$P(E \cap F) = P(F) \cdot P(E) = \frac{1}{2} \times \frac{1}{2} = \frac{1}{4}.$$

The Multiplication Rule extends to the joint occurrence of three or more events. For three events

  $$P(E \cap F \cap G) = P(F \cap G) \times P(E|F \cap G),$$

or

  $$P(E \cap F \cap G) = P(G) \times P(F|G) \times P(E|F \cap G).$$

EXAMPLE 5.

Three cards are drawn at random from an ordinary deck of 52 playing cards. What is the probability that all three are aces?

Let G be the event "an ace is drawn on the first draw," F the event "an ace is drawn on the second draw," and E the event "an ace is drawn on the third draw." Then

$$P(G) = \tfrac{4}{52} = \tfrac{1}{13}$$
$$P(F|G) = \tfrac{3}{51} = \tfrac{1}{17}$$
$$P(E|F \cap G) = \tfrac{2}{50} = \tfrac{1}{25}.$$

Therefore,

$$P(E \cap F \cap G) = \tfrac{1}{13} \times \tfrac{1}{17} \times \tfrac{1}{25} = \tfrac{1}{5,525}.$$

EXERCISES

1. Two cards are drawn at random from an ordinary deck of 52 playing cards. What is the probability that the first is a king and the second is a queen

   (a)† if the first card is replaced before the second is drawn?
   (b) if the first card is not replaced before the second is drawn?

2. Three cards are drawn at random from an ordinary deck of 52 playing cards. What is the probability that all three are 7's

   (a) if each card is replaced after it is drawn?
   (b)† if the cards drawn are not replaced?

3. A pair of dice is rolled, one red and one green. What is the probability that the red die shows a number less than 3 and the green die shows a number greater than 3?

4.† An urn contains five white and seven red balls. Two balls are drawn at random. If the first is replaced before the second is drawn, what is the probability that

   (a) two red balls are drawn?
   (b) the first ball is red and the second is white?
   (c) the first ball is white and the second is red?
   *(d) one of the balls is white and one is red?

5. What are the probabilities in Exercise 4 above if the first ball is not replaced before the second is drawn?

6. Are the two events dependent or independent in

   (a)† Exercise 1(a)?
   (b)† Exercise 1(b)?
   (c) Exercise 3?
   (d) Exercise 5?

7. Team A plays Team B in a best-of-three playoff. If the probability that Team A beats Team B in any one game is 0.40, what is the probability that Team A wins the first two games?

8.† A mathematics class consists of twenty students from the Arts and Sciences College and thirty students from the Education College. Of the arts and science students, ten are girls and ten are boys. Of the thirty education students, twenty are girls and ten are boys. A student is chosen from the class at random. Determine the probability that the student is a boy from the Arts and Sciences College

   (a)   using the Multiplication Rule.
   (b)   not using the Multiplication Rule.

9. Assume that the membership of the United States Senate is broken down as indicated.

	Democrats	Republicans	Independent
Men	45	28	2
Women	15	9	1

A senator is chosen at random to attend a White House luncheon. Determine in three different ways the probability that the one chosen is a Republican woman.

*10.† What is wrong with the following—or is it wrong?

   The probability of obtaining a 5 on the roll of a die is $\frac{1}{6}$. Therefore, the probability of obtaining two 5's on the roll of two dice is $\frac{1}{6} + \frac{1}{6} = \frac{1}{3}$.

*11. Dave and Dennis each roll a die. Use the Multiplication Rule to determine the probability that Dennis rolls a higher number than Dave. (*Hint:* Let one of the events be that the numbers rolled do not match.) Compare with Exercise 5, page 53.

12. The game of craps is played with a pair of dice. If the thrower throws a sum of 7 or 11 on the first throw, he wins. If he throws a sum of 2, 3, or 12 on the first throw, he loses. If he throws any other sum on the first throw, that sum becomes his point. He then continues to throw the dice until he obtains his point again or a sum of 7. If he throws the 7 first, he loses. If he throws his point first, he wins. What is the probability that the thrower

   (a)† wins on the first throw?
   (b)   loses on the first throw?
   (c)† throws a sum of 6 on the first throw and wins on the second throw?

(d) throws a sum of 6 on the first throw and loses on the second throw?

(e)† throws a sum of 6 on the first throw and neither wins nor loses on the second throw?

*(f) throws a sum of 6 on the first throw and wins on a subsequent throw? (*Hint:* For second event, use sample space for termination of the game.)

7.4 BINOMIAL PROBABILITIES

Suppose a die is rolled five times. What is the probability that a 4 is obtained exactly twice? The sample space for this experiment contains 6^5, or 7,776 elements. Of these, 1,250 are in the event of obtaining exactly two 4's.[1] The probability is therefore $\frac{1,250}{7,776}$, or approximately 0.16.

In this section, we will consider an alternate method of calculating probabilities for experiments, such as the above, which are repeated a number of times. First, we shall do a detailed analysis of the method applied to calculating the probability of exactly two 4's in five rolls of a die.

We shall denote the event of obtaining a 4 by S (for a success) and of obtaining a 1, 2, 3, 5, or 6 by F (for a failure). The probability of a success on any roll is $\frac{1}{6}$; the probability of a failure is $\frac{5}{6}$.

When the die is rolled five times, the success could occur on the first two rolls followed by three failures. This succession of events will be denoted by *SSFFF*. This outcome requires that each of the following events occurs:

A success on the first roll.
A success on the second roll.
A failure on the third roll.
A failure on the fourth roll.
A failure on the fifth roll.

The outcome on any roll is independent of the outcome on any of the other rolls. Therefore, the probability that each of the above events occurs is the product of their individual probabilities. That is, the probability that *SSFFF* occurs is

$$\tfrac{1}{6} \times \tfrac{1}{6} \times \tfrac{5}{6} \times \tfrac{5}{6} \times \tfrac{5}{6} = (\tfrac{1}{6})^2 (\tfrac{5}{6})^3.$$

---

1. The rolls on which the 4's appear can occur in $C(5,2) = 10$ ways; for each of these, the other three rolls can result in any of the other five numbers. Total: $10 \times 5^3 = 1,250$.

If the successes occur on the third and fifth rolls, we have *FFSFS*. As above, the probability of this outcome is

$$\tfrac{5}{6} \times \tfrac{5}{6} \times \tfrac{1}{6} \times \tfrac{5}{6} \times \tfrac{1}{6} = (\tfrac{1}{6})^2 (\tfrac{5}{6})^3.$$

Or, the successes could occur on the first and fifth rolls: *SFFFS*. The probability of this outcome is also

$$\tfrac{1}{6} \times \tfrac{5}{6} \times \tfrac{5}{6} \times \tfrac{5}{6} \times \tfrac{1}{6} = (\tfrac{1}{6})^2 (\tfrac{5}{6})^3.$$

The total number of outcomes which result in exactly two successes is $C(5,2) = 10$. Other than the three just considered, these are

SFSFF	*FSFSF*	*FFSSF*	*SFFSF*
FSFFS	*FFFSS*	*FSSFF*	

Because no two of these ten outcomes can occur together (that is, the ten are mutually exclusive) the probability that any one of them occurs is the sum of their individual probabilities. $(\tfrac{1}{6})^2 (\tfrac{5}{6})^3$ added to itself 10 times gives

$$10 \times (\tfrac{1}{6})^2 (\tfrac{5}{6})^3.$$

This product simplifies to $\tfrac{1,250}{7,776}$, the result obtained in the first paragraph of this section.

In terms of the *number of trials*, the *number of successes*, and the *probability of a success on each trial*, the probability arrived at in the last paragraph can be written

$$C(5,2)(\tfrac{1}{6})^2 (1 - \tfrac{1}{6})^3.$$

A similar analysis for the probability of just one 4 in five rolls would give

$$C(5,1)(\tfrac{1}{6})^1 (1 - \tfrac{1}{6})^4.$$

Or, for obtaining a 4 or a 5 in three out of seven rolls, we would obtain

$$C(7,3)(\tfrac{2}{6})^3 (1 - \tfrac{2}{6})^4.$$

In this last case, the probability of a success on any trial is now $\tfrac{2}{6}$, or $\tfrac{1}{3}$.

Generalizing, we would expect that the probability of obtaining r successes in n trials, when the probability of a success in each trial is p, would be given by

$$C(n,r)p^r(1 - p)^{n-r}.$$

But under what circumstances is the calculation of probabilities in this way valid?

There were two features of the above illustration that were essential in our calculations. First, the outcome on each of the five rolls was independent of the other rolls. Second, the probability of a success $(\tfrac{1}{6})$ was the same on each of the rolls. When these two conditions are not present, the above general form is not valid. (The mutual exclusiveness of the possible sequence of successes and failures will always be automatic.)

In summary, the probability of exactly r successes in n independent trials is given by

$$C(n,r)p^r(1-p)^{n-r},$$

where p is the probability of a success in each trial.

The term *success* does not refer to any special type of desirable outcome. It simply refers to the outcome of interest in a particular experiment, and is therefore a relative term. In the computation of mortality rates, a death would be considered a success!

EXAMPLE 1.

It is estimated that one out of five people have type B blood. If there are ten people in an emergency room, what is the probability that three (exactly) have type B blood, assuming that none of the ten people are related?

A success in this case is a person having type B blood. The probability of a success is $p = \frac{1}{5}$. The number of trials is $n = 10$. The probability of $r = 3$ successes is therefore

$$C(10,3)(\tfrac{1}{5})^3(\tfrac{4}{5})^7,$$

or approximately 0.201.

Note that the conditions for using the above formula are satisfied. The experiment of checking a person for blood type is considered repeated ten times. Because none of the ten people are related, the outcome on any trial is independent of the outcome on the other trials. Also, the probability of a success $(\frac{1}{5})$ remains the same on each trial.

_____

In algebra, an expression such as $p + q$, representing the sum of two quantities, is called a *binomial*. The quantities

$$C(n,r)p^r(1-p)^{n-r}$$

are called *binomial probabilities* because they are the various terms in the product

$$(p + q)^n$$

where $q = 1 - p$.

For example, if we multiplied $(p + q)$ by itself three times, we would get

$$(p + q)^3 = 1p^3 + 3p^2q + 3pq^2 + 1q^3,$$

or

$$(p + q)^3 = C(3,0)p^3q^0 + C(3,1)p^3q + C(3,2)pq^2 + C(3,3)p^0q^3.$$

The terms on the right are the probabilities of the number of successes when $n = 3$, p is the probability of a success, and $q = 1 - p$.

The number r of successes is then said to have a *binomial distribution*. Our only concern is the calculation of binomial probabilities. We mention these facts only to explain the terminology.

EXAMPLE 2.

Suppose that you are given a seven-question multiple choice quiz in which there are given four choices for each question. Because you haven't studied, you decide to randomly pick an answer to each question. What is the probability that you answer (exactly) five correctly?

Here the number of trials is $n = 7$. The probability of a success (answering correctly) on each trial is $\frac{1}{4}$. The number of successes we are interested in is $r = 5$. The probability is therefore

$$C(7,5)(\tfrac{1}{4})^5(\tfrac{3}{4})^2,$$

or approximately 0.01.

EXAMPLE 3.

Of the records produced by a certain recording company, 5% are defective. The quality control inspector randomly picks 15 records for inspection. What is the probability that he will find none defective?

A success in this case is finding a record to be defective. The probability of a success on each of the 15 trials is 0.05, or $\frac{1}{20}$. The probability of zero successes is

$$C(15,0)(\tfrac{1}{20})^0(\tfrac{19}{20})^{15},$$

or approximately 0.46.

EXAMPLE 4.

What is the probability that the inspector of Example 3 finds two or fewer defectives?

Because finding no defectives, exactly one defective, and exactly two defectives are mutually exclusive, the probability of any one occurring is the sum of their individual probabilities:

$$C(15,0)(\tfrac{1}{20})^0(\tfrac{19}{20})^{15} + C(15,1)(\tfrac{1}{20})^1(\tfrac{19}{20})^{14} + C(15,2)(\tfrac{1}{20})^2(\tfrac{19}{20})^{13}$$
$$= 0.46 + 0.37 + 0.14 = 0.97.$$

EXERCISES

Answers to the following exercises may be left in the $C(n,r)p^r(1-p)^{n-r}$ form.

1.† What is the probability of obtaining exactly three 1's on five rolls of a die?

2. In Example 1 of this section, what is the probability that exactly four of the people in the emergency room have type B blood?

3.† It is estimated that one-third of the families in the United States have two cars. If eight families are chosen at random, what is the probability that exactly four of them are two-car families?

4. If a pair of green parakeets are mated, the probability that the offspring is green is approximately 0.5. What is the probability that, among three randomly chosen offspring of green parakeets, there is exactly one green?

5.† A dart thrower in an English pub hits the bull's-eye on 80% of his shots. What is the probability that he misses the bull's-eye on exactly two of his next six shots?

6. What is the probability that a baseball player will get a hit on each of his next four times at bat if his batting percentage is 0.400?

7.† Five coins are tossed.
   (a) What is the probability that exactly three heads are obtained?
   (b) What is the probability that three or more heads are obtained?

8. Based on past experience, 25% of those who begin the Annual Joggers Mini-Marathon do not finish. On this basis, what is the probability that at least 18 of the 20 runners in this year's race will finish?

*9.† Of the tires produced by a certain company, 10% are defective. The quality control inspector randomly picks ten from each shipment. If two or more of these ten are found to be defective, the entire shipment must be examined. What is the probability that an entire shipment must be examined? (*Hint*: Consider the probability that the entire shipment will not be examined.)

7.5 PROBABILITY SUMMARY

We have seen a number of ways in which probabilities can be calculated, and a number of probabilities that have been calculated in different ways. One of the first things to be determined (and quite often is half the battle!) is the method to be used to calculate a particular probability. The methods available to us are summarized here for each reference.

1. Basic Definitions. To determine $P(E)$, it is always possible (at least theoretically) to construct the sample space, determine an appropriate probability function, determine the elements in the event E, and finally calculate $P(E)$ by adding up the probabilities assigned to the elements of E. However, this method becomes impractical if the number of elements in the sample space is very large. The remaining methods are alternatives which may be easier to use when applicable (Sections 2.3 and 7.1).

2. The Addition Rule. If E is the compound event that either of two events occur ($E = E_1 \cup E_2$), then

$$P(E) = P(E_1 \cup E_2) = P(E_1) + P(E_2) - P(E_1 \cap E_2).$$

This is particularly useful when E_1 and E_2 are mutually exclusive ($E_1 \cap E_2 = \emptyset$). Then, $P(E_1 \cap E_2) = 0$ and the Addition Rule becomes

$$P(E) = P(E_1 \cup E_2) = P(E_1) + P(E_2).$$

Use of the Addition Rule requires that $P(E_1)$ and $P(E_2)$, along with $P(E_1 \cap E_2)$ when necessary, are able to be calculated (Exercise 4, page 59 and Theorem 2, page 56).

3. The Multiplication Rule. If E is the compound event that each of two events occur ($E = E_1 \cap E_2$), then

$$P(E) = P(E_1 \cap E_2) = P(E_2) \cdot P(E_1|E_2).$$

If E_1 and E_2 are independent so that $P(E_1|E_2) = P(E_1)$, the Multiplication Rule becomes

$$P(E) = P(E_1 \cap E_2) = P(E_2) \cdot P(E_1).$$

Use of the Multiplication Rule requires that $P(E_2)$ and $P(E_1|E_2)$, or $P(E_1)$ when appropriate, are able to be calculated (Section 7.3).

4. $P(E) + P(E') = 1$**.** If $P(E')$ is known or relatively easy to calculate, then $P(E)$ can be determined from the relation

$$P(E) + P(E') = 1.$$

This applies regardless of the nature of the event E, whether E is simple or compound (Theorem 3, page 56).

5. Counting Techniques. If the elements in the sample space are equilikely, then each of the probabilities encountered above can be calculated using the counting techniques. The sample space and events do not have to be constructed. However, at least a verbal description of the sample space is required; otherwise, how could one determine whether or not the elements are equilikely (Section 7.2)?

6. Binomial Probabilities. If E is concerned with the number of successes in repeated trials of an experiment (and the other required conditions are satisfied), the easiest method for calculating $P(E)$ will often be the use of binomial probabilities (Section 7.4).

As indicated in the text, the Addition Rule and Multiplication Rule can be extended to any given number of events. This is particularly useful when working with a series of mutually exclusive or independent events.

EXERCISES

Figure 7.3 illustrates a mechanical device in which a marble is dropped into the funnel at the top and works its way through the channels into one of the cups at the bottom. At each intersection in the channels, the probability is $\frac{1}{2}$ that the marble will go in either direction.

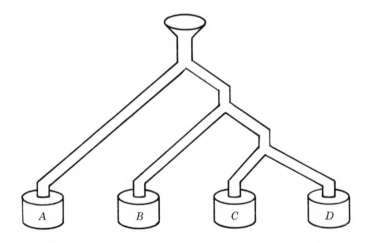

FIGURE 7.3

In the following exercises, use the method indicated to determine the probability of the given event. The number of the method corresponds to the number given in Section 7.5.

1.† Use Method 3 to determine the probability that the marble will end up in cup B; in cup C; in cup D.

2. Use Method 2 (together with Exercise 1) to determine the probability that the marble will end up in cup B, C, or D.

3.† Use Method 4 (together with Exercise 2) to determine the probability that the marble will end up in cup A.

4. Let *S1* and *S2* be sets of cups as indicated:

    S1 = {*A, B*}
    S2 = {*B, C*}.

    Use Method 2 to determine the probability that the marble will end up in a cup in either *S1* or *S2*.

5.† Set up a sample space that describes the various cups in which the marble can end its trip through the channels. Using Exercises 1 and 3, what probability should be assigned to each element of the sample space?

6. Use Method 1 and Exercise 5 to determine the probability that the marble will end up in cup *B* or in cup *C*.

7. If three marbles are dropped into the funnel, one at a time, use Method 6 to determine the probability that (exactly) two end up in

    (a)   cup *A*.   (b)† cup *B* or cup *D*.

Probability Machine. Balls dropped from the center of the top of the machine fall through the network of pins to simulate a normal distribution. (Section 8.5.)

chapter 8

statistics

Often an experiment is repeated a number of times with the results recorded and analyzed. One of the most familiar occurrences of this procedure is that of a class of students taking a test. If there are twenty students taking the test, the experiment of a student taking the test can be considered to be repeated twenty times. The twenty scores form the resulting data, sometimes referred to as the measurements. The central problem in statistics lies in determining what valid conclusions one can infer from a given set of data.

In general, statistics is concerned with collecting, describing, and interpreting data. In this chapter, we will examine some topics related to these three aspects of statistics.

8.1 ORGANIZING DATA

Usually, the "raw" data collected in an experiment is an unordered set of numbers such as:

86, 88, 93, 91, 83, 84, 84, 86, 86, 86, 89, 87, 90, 85, 91, 83,
86, 87, 93, 88, 88, 91, 90, 86, 88, 88, 90, 84, 85, 87, 85, 90,
85, 84, 85, 88, 90, 88, 87, 86, 86, 85, 84, 86, 83, 88, 86.

In this form, the numbers are practically meaningless and have to be put into a more structured arrangement for any type of analysis. There are many ways in which this can be done; we will discuss two of them.

One way is to list each number (measurement) from the smallest in the collection to the largest and to indicate the number of times that each of these appears. The number of times that each appears is called its *frequency* and the resulting table is called a *frequency table*. For the above data, the frequency table is given in Table 8.1.

TABLE 8.1

Measurement	Frequency
83	3
84	5
85	6
86	10
87	4
88	8
89	1
90	5
91	3
92	0
93	2

An alternative form in which the data can be presented is a type of bar graph called a *histogram*. A histogram for the data in Table 8.1 is given in Figure 8.1. The bottom line of the histogram lists all of the numbers from the smallest to the largest in the set of data, spaced an equal distance apart. The height of the bar centered over each measurement is equal to its frequency as found in the scale to the left.

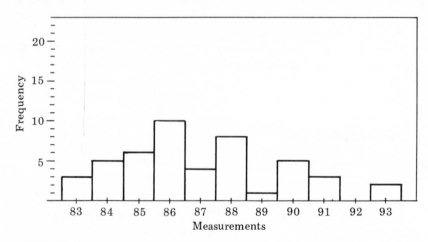

FIGURE 8.1

EXAMPLE 1.

Construct a frequency table and a histogram for the following set of data:

21.5, 23.0, 24.0, 22.5, 22.5, 23.0, 24.0, 22.0, 25.0, 24.0,
21.5, 23.0, 22.0, 22.0, 23.5, 24.0, 25.0, 24.0, 23.5, 24.0,
22.5, 23.0, 24.0, 23.0, 22.0, 22.0, 25.0, 22.0.

The frequency table is given in Table 8.2.

TABLE 8.2

Measurement	Frequency
21.5	2
22.0	6
22.5	3
23.0	5
23.5	2
24.0	7
24.5	0
25.0	3

The histogram would take the form shown in Figure 8.2.

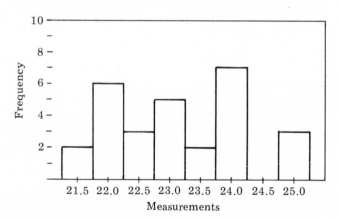

FIGURE 8.2

Note that the scales on the bottom and the left of a histogram are chosen as appropriate for a given set of data.

When the data is presented in frequency table or in histogram form, one can tell at a glance both the size of the numbers involved and their frequency. As a consequence, the data is in a form for easier analysis.

In the next sections we shall consider two types of models for describing data, as well as some uses of the two models. The two types of models are those which indicate the central tendency of the data and those which indicate the dispersion of the data. Roughly speaking, the central tendency indicates the middle of the data (in some sense), while the dispersion of the data is concerned with how the data clusters around this middle.

EXERCISES

Construct a frequency table and a histogram for each of the following sets of data.

1.† 66, 69, 69, 67, 70, 71, 72, 73, 70, 72, 66, 70, 73, 75, 73, 75, 70, 73, 71, 69, 66, 69, 72, 69, 67, 69, 70, 73, 69, 71, 66, 73, 72, 73.

2. 9.2, 9.4, 9.0, 9.4, 9.2, 9.6, 9.6, 9.8, 9.4, 9.6, 9.4, 9.6, 9.4, 9.2, 9.8, 9.6, 9.6, 10.0, 9.2, 9.4, 9.0, 9.4, 9.8, 10.0, 9.6, 9.8, 9.6.

8.2 MEASURES OF CENTRAL TENDENCY

Models that serve as measures of central tendency are called *averages*. We shall consider the two most common types of averages—the *mean* and the *median*.

The *mean* of the set of numbers $\{x_1, x_2, \ldots, x_n\}$ is the quotient

$$\frac{x_1 + x_2 + \cdots + x_n}{n}.$$

The *median* of the set of numbers $\{x_1, x_2, \ldots, x_n\}$ is the middle measurement when the numbers are arranged in order of magnitude.

EXAMPLE 1.

The number of customer calls made by a salesperson during nine consecutive weeks are 44, 47, 52, 57, 63, 48, 60, 61, and 54. The mean number of calls per week during the nine-week period is

$$\frac{44 + 47 + 52 + 63 + 48 + 60 + 61 + 54 + 57}{9} = \frac{486}{9} = 54.$$

If the number of calls are arranged in increasing order

44, 47, 48, 52, ⑤④, 57, 60, 61, 63,

it is seen that the median number of calls is also 54.

EXAMPLE 2.

Suppose that the number of calls made by the salesperson of Example 1 during the next nine weeks are

21, 35, 37, 48, (54), 55, 59, 61, 62.

The median is again 54, but the mean is

$$\frac{21 + 35 + 37 + 48 + 54 + 55 + 59 + 61 + 62}{9} = 48.$$

The medians in these two examples are the same but the means differ. The smaller mean in Example 2 is due to the fact that extreme values, like the relatively small values 21, 35, and 37, influence a mean strongly. The following example again illustrates this possibility.

EXAMPLE 3.

Suppose that the years of service of the 24 teachers in a particular high school are 1, 1, 2, 2, 2, 2, 3, 3, 3, 3, 3, 3, 4, 4, 5, 5, 6, 6, 6, 6, 9, 10, 25, and 30. The mean number of years of service of these teachers is 6. Actually 16, or two-thirds, of the teachers had less than six years of service. This is due to the relatively high number of years of service of the two teachers with 25 and 30 years.

In this situation, the median of 3.5 years of service would indicate that half of the teachers had less than 3.5 years of service, despite the fact that the mean number of years is 6!

Note that the number of teachers is an even number; there is no middle measurement. The median of such a set of numbers is defined to be the mean of the two middle numbers. In this example, the twelfth largest number is 3 and the thirteenth number is 4. The median is

$$\frac{3 + 4}{2} = 3.5.$$

Both the mean and the median are mathematical descriptions (models) of sets of data. Each, although in different ways, answers the question, "What is the central tendency of the data?" and the result is interpreted in the physical situation as the middle measure of the number of salesperson calls (Example 1), the middle measure of years of service (Example 3), and so forth. Several uses of these middle measures are discussed in the following paragraphs.

A common use of the averages is to represent a set of data by a single number. In using the averages in this way, it is preferable to indicate both the median and the mean. If only one is to be used, it should be remembered that the mean divides the total evenly among the number of entries; the median indicates the middle position of the entries.

If both the mean and the median are indicated, a mean that is much greater than the median would suggest the possibility of a few numbers that are considerably larger than the others. Conversely, a mean that is much smaller than the median would suggest a few numbers considerably smaller. Remember that the mean and median are single numbers representing a whole set of numbers. Detail is sacrificed for the sake of simplicity. Whether this is misleading depends on the particular purpose. In some cases, it may be more desirable to examine the whole set of numbers.

Besides representing sets of data by single numbers, averages are used to compare the relative size of one of the numbers to the total set. The expressions "above average" and "below average" are often used in this connection.

One of the most common occurrences of this second use of averages is in connection with a class taking a test. After the results are known, a student compares his score with the class mean to determine how he did relative to the rest of the class. Or, he compares his score with the class median to determine whether or not he scored in the top half of his class.

A third use of averages is the comparison of two sets of data. The average monthly maintenance cost of several automobiles might be compared to determine which is the most economical. The average score of two classes taking the same exam might be compared to determine which class is superior. However, the use of averages for the sake of comparison of two groups of people or objects can be misleading. The possibility of error in this use of averages will be examined further in Section 8.4.

If the set of data is large and the data entries are repeated several times, as was the case in the last section, the calculation of the mean can be simplified by using the frequencies. Instead of adding each measurement the number of times it appears, the measurements are multiplied by their frequencies and these products are added. The resulting sum is the same, and the calculations are generally simplified. This procedure is illustrated in the following example.

EXAMPLE 4.

Calculate the mean of the set of data given at the beginning of Section 8.1.

The measurements, their frequencies, and the required products are given in Table 8.3.

TABLE 8.3

Measurement	Frequency	Frequency X Measurement
83	3	249
84	5	420
85	6	510
86	10	860
87	4	348
88	8	704
89	1	89
90	5	450
91	3	273
93	2	186
Total:	47	4,089

The mean is then 4,089/47 = 87.

Two other measurements of central tendency that are sometimes used, but not as frequently as the two we have considered, are the mode and the midrange. The *mode* is the measurement that occurs most often. For the data in Table 8.3, the mode is 86, because it has the highest frequency.

The *midrange* is the mean of the largest and smallest measurement. For the data in Table 8.3, the midrange is

$$\frac{83 + 93}{2} = \frac{176}{2} = 88.$$

Neither the mode nor the midrange will be considered further in this text.

EXERCISES

1. Determine the mean and the median for the sets of numbers:

   (a)† 19, 23, 20, 34, 31, 35, 30, 26, 25.
   (b) 199, 204, 185, 180, 190, 210, 220, 235, 216, 221.

2. Two types of fertilizer are tested for yield of corn on 15 pairs of plots of soil. The yield in bushels is given in Table 8.4.

   (a) Determine the mean yield for each type of fertilizer.
   (b) On the basis of the means, which type of fertilizer is better?
   (c) Determine the median yield for each type of fertilizer.
   (d) On the basis of the medians, which type of fertilizer is better?

TABLE 8.4

Type A	Type B	Type A	Type B
46.3	47.9	43.7	42.4
42.1	40.3	42.0	41.3
40.1	39.6	44.0	45.1
38.3	42.6	47.2	42.2
39.2	29.9	44.3	44.1
41.9	40.6	39.8	40.2
39.2	38.1	41.9	41.9
		40.1	41.9

3.† Two brands of tires were tested simultaneously on the rear wheels of eleven cars. Table 8.5 indicates (in thousands) the number of miles each car traveled before the tread was reduced to a certain level.

TABLE 8.5

Type I	Type II	Type I	Type II
40.2	40.1	42.4	41.4
42.3	41.0	39.9	39.7
41.6	42.1	42.8	42.7
43.7	42.5	43.1	42.5
41.9	41.1	44.1	43.2
		43.8	42.8

(a) Determine the mean number of miles traveled for each type of tire.

(b) On the basis of the means, which type of tire is better?

(c) Determine the median number of miles traveled for each type of tire.

(d) On the basis of the medians, which type of tire is better?

 4. Calculate the mean and the median of the set of data in Exercise 1, page 244.

 5.† Calculate the mean and the median of the set of data in Exercise 2, page 244.

*6. Construct a set of nine numbers that has the same number for its mean, median, and mode. (*Hint:* Start in the "middle.")

8.3 MEASURES OF DISPERSION

Models for analyzing the dispersion of data are intended to indicate how closely the data centers about an average. Compare the histogram in Figure 8.3 on page 249 with that given in Figure 8.1, page 242. Both represent

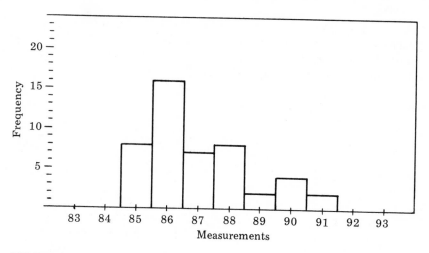

FIGURE 8.3

47 measurements; the corresponding sets of data each have a mean of 87 and a median of 86. In this latter case, the data clusters more closely about the averages. Correspondingly, the measures of dispersion (or of spread) would be smaller in this latter case than in the former.

The standard deviation measures the closeness of a set of numbers about their mean. If one wanted to measure this closeness, a natural approach would be to calculate the difference between each number and the mean, and then determine the mean of these differences. But there is a problem involved in this approach! Consider the set

{17, 18, 21, 25, 28, 29, 30, 32}

with mean 25. If we calculate the differences and add them, we get:

$$17 - 25 = -8$$
$$18 - 25 = -7$$
$$21 - 25 = -4$$
$$25 - 25 = 0$$
$$28 - 25 = 3$$
$$29 - 25 = 4$$
$$30 - 25 = 5$$
$$32 - 25 = 7$$
$$\overline{0}$$

The mean of the differences would be $\frac{0}{8} = 0$. For any set of numbers, we would always get the same result. Because of the nature of the mean, the positive differences would always be offset by corresponding negative differences.

Instead, the standard deviation is based on the mean of the *square* of the differences. The square root of this mean gives the standard deviation. Formally, if $\bar{x}$ denotes the mean of the set of numbers $\{x_1, x_2, \ldots, x_n\}$, the *standard deviation* is the square root of the quotient

$$\frac{(x_1 - \bar{x})^2 + (x_2 - \bar{x})^2 + \cdots + (x_n - \bar{x})^2}{n}.$$

The notation for the numerator of the above quotient is

$$\sum_{i=1}^{n} (x_i - \bar{x})^2.$$

The letter i is called the index of summation; the notation indicates that i should take on the values $1, 2, 3, \ldots, n$ in the expression $(x_i - \bar{x})^2$ and the results added. With this notation, the standard deviation can be written

$$\sqrt{\frac{\sum_{i=1}^{n} (x_i - \bar{x})^2}{n}}.$$

EXAMPLE 1.

Determine the standard deviation of the set of numbers

$$\{17, 18, 21, 25, 28, 29, 30, 32\}.$$

The mean of the set of numbers is 25.

$$(17 - 25)^2 = (-8)^2 = 64$$
$$(18 - 25)^2 = (-7)^2 = 49$$
$$(21 - 25)^2 = (-4)^2 = 16$$
$$(25 - 25)^2 = 0^2 = 0$$
$$(28 - 25)^2 = 3^2 = 9$$
$$(29 - 25)^2 = 4^2 = 16$$
$$(30 - 25)^2 = 5^2 = 25$$
$$(32 - 25)^2 = 7^2 = 49$$

$$\sum_{i=1}^{8} (x_i - 25)^2 = 64 + 49 + 16 + 0 + 9 + 16 + 25 + 49 = 228.$$

The standard deviation is therefore

$$\sqrt{\frac{228}{8}} = \sqrt{28.5},$$

which is approximately 5.3.

If a procedure (such as a hand calculator!) is not readily available for determining the value of a square root such as $\sqrt{28.5}$, an approximate value can be found using a procedure similar to the following.

The closest perfect squares to 28.5 are 25 and 36 so that

$$5 < \sqrt{28.5} < 6.$$

Moreover, 28.5 is about one-third the distance from 25 to 36:

We then take our estimate of $\sqrt{28.5}$ to be about one-third the distance from 5 to 6, or 5.3.

This estimate must be tested by squaring it:

$$
\begin{array}{r}
5.3 \\
5.3 \\
\hline
15\ 9 \\
265 \\
\hline
28.09
\end{array}
$$

Because 28.09 is reasonably close to 28.5, 5.3 is a reasonable approximation for $\sqrt{28.5}$.

If the result of the squaring had not been close to 28.5, a higher or lower estimate would have to be chosen and tested until a satisfactory approximation is obtained.

EXAMPLE 2.

The set of numbers in this example has the same mean as does the set in Example 1. Determine the standard deviation and compare with that of Example 1.

$$\{1, 7, 12, 25, 28, 35, 43, 49\}$$

$$
\begin{aligned}
(1 - 25)^2 &= (-24)^2 = 576 \\
(7 - 25)^2 &= (-18)^2 = 324 \\
(12 - 25)^2 &= (-13)^2 = 169 \\
(25 - 25)^2 &= \quad 0^2 \ \ = \quad 0 \\
(28 - 25)^2 &= \quad 3^2 \ \ = \quad 9 \\
(35 - 25)^2 &= \quad 10^2 = 100 \\
(43 - 25)^2 &= \quad 18^2 = 324 \\
(49 - 25)^2 &= \quad 24^2 = 576
\end{aligned}
$$

$$\sum_{i=1}^{8} (x_i - 25)^2 = 2{,}078$$

The standard deviation is therefore

$$\sqrt{\frac{2,078}{8}} = \sqrt{259.75},$$

which is approximately 16.1.

The greater standard deviation of this example as compared with that of Example 1 is accounted for by the greater distances of the numbers from their respective means.

The position of one of the numbers of a set of data relative to the whole set can be indicated by not only whether the number is above or below the mean, but more exactly by determining how many standard deviations above or below.

EXAMPLE 3.

Suppose the numbers of the set of data of Example 1 are the scores made by a class of eight students on a 35-question true-false examination. Betty, one of the eight students, had the score of 30. How did she do relative to the rest of the class?

Betty's score of 30 indicates that she scored five points above the mean. Her score is 5/5.3, or approximately one, standard deviation above the mean.

EXAMPLE 4.

Suppose that the numbers in the set of data of Example 2 are the scores that the same class of eight students as in Example 3 scored on a subsequent true-false examination of 50 questions. On this test, Betty scored 35. Did she do better on the second test than she did on the first, relative to the whole class?

Because she scored ten points above the mean on the second test, she scored 10/16.1, or approximately 5/8, standard deviation above the mean. As compared to the one standard deviation above the mean of the first test, 5/8 standard deviation above the mean of the second test indicates that Betty did worse on the second test relative to the entire class.

Just as a set of data is divided into halves by the median, it is divided into fourths by the quartiles. These, like the median, are positions determined by counting. The *first quartile* (Q_1) of the set of data $\{x_1, x_2, \ldots, x_n\}$ is a number such that one-fourth of the measurements are less than or equal to this number. The *third quartile* (Q_3) is a number such that one-fourth of the measurements are greater than or equal to this number. The *second quartile* is the median itself.

The difference between the third and first quartiles, $Q_3 - Q_1$, is called the *interquartile range* and measures the spread of the data about the median.

EXAMPLE 5.

Determine the interquartile range of the set of numbers of Example 1:

$\{17, 18, 21, 25, 28, 29, 30, 32\}$.

Because there are eight numbers in the set and because $\frac{1}{4} \times 8 = 2$, the first quartile should be the second number (because they are already given in increasing order). Q_1 is therefore 18.

Similarly, Q_3 should be the second-to-last number, or 30. The interquartile range is $Q_3 - Q_1 = 30 - 18$, or 12.

EXAMPLE 6.

The set of numbers of Example 2,

$\{1, 7, 12, 25, 28, 35, 43, 49\}$,

has the same median, 26.5, as does the set of Example 5. Determine its interquartile range and compare with that of Example 5.

Here, $Q_1 = 7$ and $Q_3 = 43$, so that the interquartile range is $43 - 7 = 36$. The much larger result here as compared with that of 12 in Example 5 reflects the fact that the numbers in this case are much more spread out than in the former.

The position of a number in a set of data, when the median is used as an average, can be indicated not only by determining whether the number is above the median but also whether the number is above or below the third quartile. If the number is below the median, it is more informative to know also whether the number is above or below the first quartile.

Betty, of Examples 3 and 4, scored above the median in each test—at the third quartile on the first, below on the second. On this basis as well, Betty did better on the first test relative to the entire class.

Some common uses of the standard deviation and quartiles have been indicated in the examples of this section:

1. The spread of data about the mean or median, as in Examples 1 and 5.
2. A comparison of the spread of the data about the mean or median of two or more sets of data, as in Examples 2 and 6.
3. The relative position of one member of a set of data, as in Example 3.
4. The relative positions of corresponding members of two or more sets of data are compared, as in Example 4 and in the discussion of the last paragraph.

As an illustration of the application of mathematics, the measures of dispersion provide mathematical models to answer questions of how widely the data is spread about the mean or median. Calculations are within the set of real numbers. The results are interpreted in the manner indicated in the previous examples.

Just as in the calculation of the mean, the calculation of the standard deviation and quartiles can be simplified by use of frequencies when the set of data is large and the measurements are repeated a number of times. In calculating the standard deviation, the square of the difference for each measurement is multiplied by its frequency and the results are added, as illustrated in the following example.

EXAMPLE 7.

Calculate the standard deviation of the set of data given in Table 8.3.

The mean of the set of data is 87. The quantities required for the calculation of the standard deviation are given in Table 8.6.

TABLE 8.6

Measurement	Frequency	Frequency Times Square of Distance from Mean
83	3	3 × 16 = 48
84	5	5 × 9 = 45
85	6	6 × 4 = 24
86	10	10 × 1 = 10
87	4	4 × 0 = 0
88	8	8 × 1 = 8
89	1	1 × 4 = 4
90	5	5 × 9 = 45
91	3	3 × 16 = 48
93	2	2 × 36 = 72
Total:	47	Sum: 304

The standard deviation is

$$\sqrt{\frac{304}{47}},$$

or approximately 2.5.

Because there are 47 measurements in the set of data of this last example, the determination of the quartiles would result in the fraction

$$\tfrac{1}{4} \times 47 = 11\tfrac{3}{4}.$$

In such situations, the fraction is rounded to the nearest integer. In this case, Q_1 would be the twelfth number from the bottom, or 85. Q_3 would be the twelfth from the top, or 88. The interquartile range is therefore $88 - 85 = 3$.

The one exception to the rule just given for calculating quartiles occurs when the fraction involved is $\frac{1}{2}$. In this case, the mean of the numbers above and below is used. For example, the set

$\{2, 7, 8, 12, 15, 20, 22, 25, 27, 30\}$

has ten measurements. Because $\frac{1}{4} \times 10 = 2.5$, the first quartile is given by the mean of the second and third measurements:

$$Q_1 = \frac{7 + 8}{2} = 7.5.$$

Similarly, the third quartile is given by the mean of the second and third largest measurements:

$$Q_3 = \frac{25 + 27}{2} = 26.$$

You are probably more familiar with the term *percentile* than with *quartile*. Percentiles divide a set of data into hundredths instead of fourths. One can define an interpercentile range and locate a particular number in a set of data using percentiles in much the same way as is done with quartiles. You have probably seen this done in the interpretation of standardized tests. However, the sets of data with which we are concerned in this text are not large enough to be divided into hundredths.

A final measure of dispersion for a set of data is the difference between its largest and smallest numbers. This difference is called its *range*. For the set just considered,

$\{2, 7, 8, 12, 15, 20, 22, 25, 27, 30\}$,

the range is $30 - 2 = 28$.

The range measures the spread of the data independent of any measure of central tendency or average.

EXERCISES

1.† Determine the standard deviation, the interquartile range, and the range of the measurements

   $82, 62, 81, 63, 80, 64, 66, 78, 68, 75, 73, 72.$

2. Determine the standard deviation, the interquartile range, and the range of the measurements

   $87, 86, 85, 85, 84, 81, 78, 77, 75, 75, 74, 73.$

3.† Suppose that the data of Exercises 1 and 2 are the scores that a class of 12 students made on two different tests. If Joan scored 66 on the first test (Exercise 1) and 74 on the second,

   (a) Was Joan's score within one standard deviation of the mean on the first test? On the second test?
   (b) On the basis of part (a), did Joan do better on the first or on the second test, relative to the entire class?

4. Suppose that grades are given on the following basis:

   A — scores above 2 standard deviations above the mean.
   B — scores above 1, but less than 2, standard deviations above the mean.
   C — scores in the range from 1 standard deviation below to 1 standard deviation above the mean.
   D — scores less than 1, but greater than 2, standard deviations below the mean.
   F — scores less than 2 standard deviations below the mean.

   On this basis, what grade did Joan of Exercise 3 get on the first test? On the second test?

5.† Calculate the standard deviation, the interquartile range, and the range of the set of data given in Exercise 1, page 244.

6. Calculate the standard deviation, the interquartile range, and the range of the set of data given in Exercise 2, page 244.

7. Verify that addition of the differences of the numbers in Exercise 1 from their mean gives a sum of 0.

8.4 THE SIGN TEST

The material of the previous sections is part of *descriptive statistics*; the object was to describe a set of collected data. As was pointed out at the beginning of this chapter, the central problem in statistics is the determination of the conclusions which validly follow from the data. This aspect of statistics is called *inferential statistics.* The purpose of this section is to examine one of the methods used in establishing a conclusion based on a set of data.

   In inferential statistics, the set of data is usually a *sample* drawn from a much larger collection, called the *population.* The object is to arrive at a conclusion concerning the total population on the basis of the sample. One of the most familiar examples of this procedure is the forecasting of the results by the television networks on election nights. Precincts representative of all of the voters in a state are selected. On the basis of the voting results in the selected precincts, the outcome is forecast for the whole state. In this case, the voters in the selected precincts form the sample; the totality of voters in the state form the population.

One of the first requirements, of course, is that the sample be truly representative of the whole population. The study of correct procedures for choosing a sample is the subject of a branch of statistics called *sampling theory*. We will not go into this aspect of statistics here.

Suppose that Table 8.7 gives the number of items produced per day by two machines operated by the same individual during consecutive three-week periods. The given data is the sample; the daily production during the lives of the respective machines is the population.

TABLE 8.7

Machine A	Machine B
880	930
930	940
920	890
870	890
850	900
870	880
950	930
890	910
910	940
890	930
900	910
920	900
890	920
900	930
870	880
Mean 896	Mean 912
Median 890	Median 910

The higher mean and median of Machine B would indicate that Machine B is to be preferred. But is this difference due to chance? If three different weeks were chosen to compare the machines, would the results have indicated that Machine A is to be preferred to B?

It is the company's intention to purchase the better of the two machines; both cost about the same. The operator maintains that Machine B is in fact the better of the two and recommends it for purchase. His supervisor, who has been wined and dined by the sales representative for Machine A, maintains otherwise. He maintains that the difference in the trial runs is due to chance, and that, if the trial runs could be extended, the results might well be reversed.

However, the agreement with each manufacturer was for a three-week trial period. The management consults their statistician to determine whether or not the data in Table 8.7 indicate a real difference in the long-range performance of the two machines. The statistician bases her decision on the following sign test.

The difference in the production of the two machines for each day is determined:

$$880 - 930 = -50 \quad -$$
$$930 - 940 = -10 \quad -$$
$$920 - 890 = 30 \quad +$$
$$870 - 890 = -20 \quad -$$
$$850 - 900 = -50 \quad -$$
$$870 - 880 = -10 \quad -$$
$$950 - 930 = 20 \quad +$$
$$890 - 910 = -20 \quad -$$
$$910 - 940 = -30 \quad -$$
$$890 - 930 = -40 \quad -$$
$$900 - 910 = -10 \quad -$$
$$920 - 900 = 20 \quad +$$
$$890 - 920 = -30 \quad -$$
$$900 - 930 = -30 \quad -$$
$$870 - 880 = -10 \quad -$$

The "+" sign indicates that the corresponding difference is greater than 0; the "−" sign indicates that the corresponding difference is less than 0. If there is no real difference in the output of the two machines, the probability of a plus or a minus on any one day ought to be the same; that is, each should have a probability of $\frac{1}{2}$. In this case, there are 12 minus signs and 3 plus signs. The sign test considers the larger of the two—in this case the 12 minus signs. What is the probability of 12 or more minus signs when the probability of a minus on any one day is $\frac{1}{2}$?

In a 15-day period,

the probability of 12 minus signs is 0.0139,[1]
the probability of 13 minus signs is 0.0032,
the probability of 14 minus signs is 0.0005, and
the probability of 15 minus signs is 0.0001.

Because getting (exactly) 12, 13, 14, or 15 minus signs are mutually exclusive, the probability of getting 12 or more is their sum

$$0.0139 + 0.0032 + 0.0005 + 0.0001 = 0.0177.$$

Because the probability of obtaining a value in this range (12 to 15) is so small when minus has a probability of $\frac{1}{2}$, we can conclude that minus actually has a probability greater than $\frac{1}{2}$. But if minus, in fact, does have a

[1]These probabilities are calculated using binomial probabilities of Section 7.4. However, we will not have to calculate these probabilities to use the sign test.

probability greater than $\frac{1}{2}$, then Machine B can be expected to have the greater daily output over the lives of the two machines!

In general, the sign test begins with the assumption that there is no difference in the objects being compared (the daily performance of the two machines in the above illustration), so that plus and minus each have a probability of $\frac{1}{2}$. Then, one decides to accept or reject this assumption, depending upon the number of plus signs and minus signs obtained in the sample. If the greater of these two numbers lies in a range with very small probability, the assumption of no difference is rejected; otherwise, the assumption is accepted. If the assumption is rejected, then one of the objects is considered to be better than the other.

Two questions arise. How small is a "small probability," and how much confidence can we have in our decision? We will return to these questions after some further illustrations.

EXAMPLE 1.

A company has developed an additive for its insect repellent and tested the repellent both with and without the additive in 14 different houses. A record was kept of the number of days each was effective; this record is given in Table 8.8.

TABLE 8.8

With	Without	With	Without
139	130	158	151
164	160	148	144
145	138	160	157
160	170	130	133
183	183	148	142
151	159	150	160
149	145	164	171

Use the sign test to determine whether the data indicate that the repellent with the additive is better.

Taking the differences of the data for each of the 14 houses above yields the following table.

+	+
+	+
+	+
−	−
0	+
−	−
+	−

There are 8 plus signs and 5 minus signs; a difference of 0 is ignored. What significance should we attach to the 8 plus signs, the greater of the two numbers?

We assume first that there is no difference between the repellent with the additive and the repellent without the additive. Then, the probability of plus or minus at any house is $\frac{1}{2}$. Suppose that we will reject the assumption of no difference only if the probability of 8 or more plus signs is less than 0.05.

In a total of 13 houses,

> the probability of 13 plus signs is 0.0001,
> the probability of 12 plus signs is 0.0016,
> the probability of 11 plus signs is 0.0095,
> the probability of 10 plus signs is 0.0349,
> the probability of   9 plus signs is 0.0873, and
> the probability of   8 plus signs is 0.1571.

Then the probability of 12 or 13 plus signs is $0.0016 + 0.0001 = 0.0017$, the probability of 11, 12, or 13 plus signs is $0.0095 + 0.0016 + 0.0001 = 0.0112$, the probability of 10, 11, 12, or 13 plus signs is $0.0349 + 0.0095 + 0.0016 + 0.0001 = 0.0461$, and the probability of 9, 10, 11, 12, or 13 plus signs is $0.0873 + 0.0349 + 0.0095 + 0.0016 + 0.0001 = 0.1334$.

In the same way, the probability of 8, 9, 10, 11, 12, or 13 plus signs is the sum of their individual probabilities, or 0.2905.

Because we decided to reject the assumption of no difference only if this last probability was less than 0.05, we accept the assumption of no difference and conclude that the additive does not improve the effectiveness of the repellent.

Note carefully in Example 1 that the probability of 10 or more plus signs is less than 0.05 and the probability of 9 or more is greater than 0.05 —that is, the transition from less than 0.05 to greater than 0.05 is made when going from 10 or more to 9 or more. The smallest number of plus signs for which we would have rejected the assumption of no difference is 10. Figure 8.4 indicates the number of plus signs, or of minus signs, as the case may be, for which the assumption of no difference is rejected and those for which it is accepted when 13 comparisons are counted.

The smallest number in a rejection region is called the *critical value*. The critical value varies with the number of comparisons counted; in Figure 8.4, the critical value is 10.

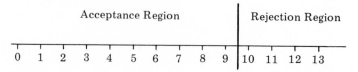

FIGURE 8.4

In Table 8.9, n denotes the number of comparisons counted, that is, the sum of the number of plus signs and the number of minus signs. (Remember that zeros are ignored.) Table 8.9 gives the critical value for n up to 47.

TABLE 8.9 Critical Values for the Sign Test (0.05 Level)

n	Critical Value	n	Critical Value	n	Critical Value
6	6	20	15	34	23
7	7	21	15	35	23
8	7	22	16	36	24
9	8	23	16	37	24
10	9	24	17	38	25
11	9	25	18	39	26
12	10	26	18	40	26
13	10	27	19	41	27
14	11	28	19	42	27
15	12	29	20	43	28
16	12	30	20	44	28
17	13	31	21	45	29
18	13	32	22	46	30
19	14	33	22	47	30

To use the sign test, all that is required is to determine whether the number of plus signs or the number of minus signs, whichever is the greater, is less than the appropriate critical value for acceptance, or greater than or equal to the critical value for rejection.

In Example 1, $n = 8 + 5 = 13$. The greater number of signs is the 8 plus signs and the critical value is 10. Because 8 is less than 10, the assumption of no difference is accepted.

EXAMPLE 2.

Two classes of 34 students each are taught mathematics by one of two different methods. The students are paired by IQ. Taking difference of scores yields 9 minus signs and 25 plus signs. Should the assumption of no difference in the methods be accepted or rejected?

The greater number of signs is the 25 plus signs. The value of n is $9 + 25 = 34$. For $n = 34$, the critical value in Table 8.9 is 23. Because there are more than 23 plus signs, we reject the assumption that there is no difference in the results of the two methods. The data indicate that one method is superior to the other.

EXAMPLE 3.

In Example 2, suppose that taking difference of scores yields 18 minus signs and 16 plus signs. Should the assumption of no difference in the methods be accepted or rejected?

Because the greater number of signs is 18, which is less than the critical value of 23, the assumption of no difference in methods is accepted.

_____

The sign test is applicable whenever the entries being compared in the sample would be expected to be the same if there were no difference in the objects being evaluated. In Example 2 above, the students are paired by IQ; if they were paired by sex or the two classes had different instructors, the difference in scores might not indicate anything as far as the superiority of either method is concerned. Similarly, the two machines in the illustration would have to be treated as equally as possible, being installed in identical circumstances, being operated by the same individual, and the time periods being fairly chosen. If, for example, they had been operated by different individuals for three-week periods, the difference in output might have been due to the difference in efficiency of the operators and not to the difference in machines.

We return now to the question of what confidence we can have in a decision made in this way. When using Table 8.9, note that the probability of obtaining a value in the critical region when the assumption of no difference is true is 0.05. Consequently, the probability of rejecting this assumption when, in fact, it is true is 0.05. That is, an error of this type can be expected to occur about 5% of the time in the long run. The *size* of the critical region is said to be 0.05.

EXAMPLE 4.

Thirty-five students each perform the following experiment with the same coin to determine if the coin is biased.

The coin is tossed forty times by each student. Each keeps a record of his or her results by recording a "1" under heads and a "0" under tails whenever a head is obtained, and vice versa when a tail is obtained. The following record indicates two heads followed by a tail.

Heads	*Tails*
1	0
1	0
0	1
.	.
.	.
.	.

When the differences are taken, a plus corresponds to obtaining a head; a minus corresponds to obtaining a tail.

(The coin is known to be unbiased, but the students are not told this.)

Thirty-four of the students obtain a greater number of plus signs or minus signs in the 20 to 25 range. Because the critical value for $n = 40$

is 26, each of these thirty-four students accept the assumption of no difference in the number of heads and tails to be expected in the long run.

The thirty-fifth student obtains 27 minus signs and 13 plus signs. He therefore rejects the above assumption!

But note that all thirty-five students have made the *statistically correct* decision. The last one happens to be erroneous. He rejected an assumption which happened to be true.

We can vary the likelihood of this type of error by varying the size of the critical region. Some common sizes are 0.01, 0.025, 0.05, and 0.10. The risk of the type of error we have been discussing changes accordingly. For Table 8.9, the size 0.05 was chosen arbitrarily, except for the fact that it is one of those that are commonly used.

There is also the possibility of accepting the assumption of no difference when, in fact, it is false. If the coin in Example 4 had been known to be biased, the thirty-four students would have made this type of error— even though their decision was again statistically correct.

An analysis of this second type of error is beyond the scope of this text. In practice, the statistician chooses the size of the critical region according to the risk he or she is willing to take of committing each of the two types of error indicated, and the relative importance of either type of error in a particular application.

EXERCISES

In each of the following exercises use the sign test with critical region size 0.05.

1.[†] Table 8.10 gives the miles per gallon obtained by 15 drivers using two types of spark plugs. Determine if the results indicate either type of plug to be superior.

TABLE 8.10

Type A	Type B	Type A	Type B
14.3	13.9	19.3	18.8
16.7	16.5	14.9	14.7
15.3	14.9	15.8	15.3
16.0	16.9	16.2	16.2
14.9	14.3	13.8	13.1
15.4	16.0	20.1	18.9
17.3	16.9	18.3	17.5
20.3	20.1		

2. Table 8.11 gives the weights of 17 people before and after they participated in a one-month weight reducing program. Determine whether the data indicate that the plan is effective.

TABLE 8.11

Before	After	Before	After	Before	After
144	138	176	180	151	148
192	190	222	212	158	161
139	130	150	146	138	141
193	193	176	174	132	130
160	164	120	130	211	202
149	145	170	166		

3.† Determine if either of the fertilizers in Table 8.4, page 248, is superior.

4. Determine if either type of tire in Table 8.5, page 248, is superior.

5.† Forty-five people chosen at random tasted two different brands of chianti wine. Of the forty-five,

   27 preferred the domestic brand,
   15 preferred the imported brand, and
   3 had no preference.

Determine whether the results indicate that the domestic brand is more appealing to the general public.

8.5 THE NORMAL CURVE

When a coin is tossed five times, the number of heads that could be obtained and their respective probabilities are given in Table 8.12.

TABLE 8.12

Number of Heads	Probability[1]
0	0.03
1	0.16
2	0.31
3	0.31
4	0.16
5	0.03

---

[1] These probabilities are calculated using binomial probabilities of Section 7.4, with $p = \frac{1}{2}$. However, we will not have to calculate these probabilities to use the normal curve.

To consider the graphic effect of this information, we construct a bar graph, similar to a histogram (Fig. 8.5). Now, however, the bar over each measurement represents the probability of obtaining that measurement.

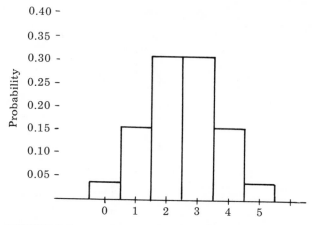

FIGURE 8.5

The sum of the probabilities represented is 1. Therefore, we consider the total area of the graph to be one unit. The area of the bar over each measurement is equal to the probability of obtaining that particular measurement.

We now connect the midpoints of the top of the bars with a polygonal line as indicated in Figure 8.6. The probability of obtaining any particular number of heads is now approximately the same as the corresponding area under the line. Also the probability of obtaining from, say two to four heads would be approximately the same as the area under the polygonal line above the intervals corresponding to 2, 3, and 4.

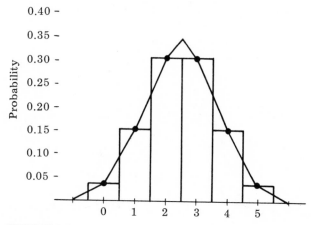

FIGURE 8.6

As the number of tosses increases, these approximations become better. When the coin is tossed nine times, the resulting polygonal line is indicated in Figure 8.7. When the number of tosses is very large and the

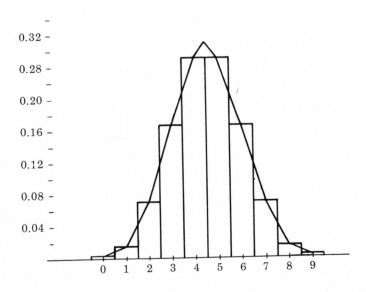

FIGURE 8.7

base of the figure is maintained at the same width, the bars become essentially lines over the different measurements. At the same time, the area approximations based on the polygonal lines for obtaining a number of heads in any range, say from 75 to 100 heads in 200 tosses, is very good.

Also, as the number of tosses becomes very large, the polygonal lines take on the appearance of smooth curves in the shape of a *bell-shaped* or *normal* curve as in Figure 8.8. While there are many types of curves used

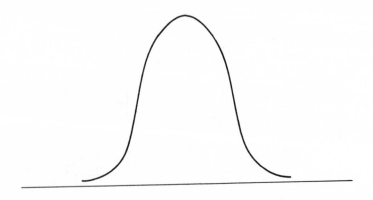

FIGURE 8.8

in statistics, the normal curve is by far the most important. Its use as a model for the distribution of so many various quantities is consistent with its title—the normal curve.

The determination of probabilities using the normal curve is also based on the area under the curve. The total area under the curve is one unit, and the probability of obtaining a value between two numbers, *a* and *b*, is the same as the area under the curve indicated in Figure 8.9. Quantities for which the probabilities are determined in this manner are said to have a *normal distribution*. Examples of quantities having a normal distribution will be given in the illustrations and exercises of this section. Our immediate objective is to examine the characteristics of the normal curve and the process of determining probabilities using this curve.

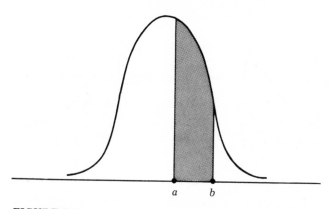

a b

FIGURE 8.9

Distributions such as the normal also have a mean and standard deviation. The manner in which these are defined and calculated is entirely different from those for sets of data. However, their purposes are similar— to indicate the center of the distribution and the spread of the distribution about the center. We will use the standard notation: the Greek letters mu (μ) for the mean and sigma (σ) for the standard deviation.

We will not examine the manner in which the mean and standard deviation of a normal distribution are calculated, as these are fairly complicated. It is sufficient for our purposes to realize these two characteristics:

1. The mean is the balance point of the distribution. Half of the area under the curve lies to the right of the mean and half to the left (Fig. 8.10). It is also that value at which the normal curve peaks.
2. The standard deviation is so chosen that the area under the curve is distributed approximately in the manner indicated in Figure 8.11.

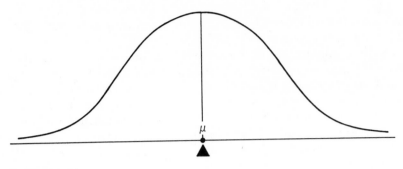

FIGURE 8.10

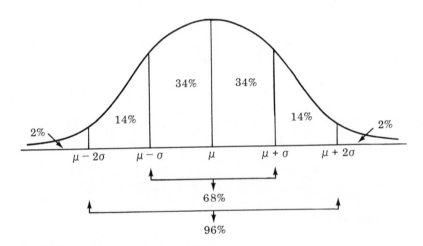

FIGURE 8.11

If the possible outcomes of an experiment have a normal distribution, the probability of obtaining an outcome within one standard deviation of the mean is therefore 0.68; the probability of obtaining an outcome within two standard deviations is 0.96. For example, suppose that the life of a particular type of washing machine is normally distributed with a mean of 5 years and standard deviation of 6 months. Then, 68% of these machines have a life from $4\frac{1}{2}$ to $5\frac{1}{2}$ years, and the probability that any one of these machines will have a life span from $4\frac{1}{2}$ to $5\frac{1}{2}$ years is 0.68.

Figure 8.12(a) depicts several normal curves that have the same mean, but different standard deviations. In Figure 8.12(b), these relationships are reversed; the curves have the same standard deviation but different means.

In Figure 8.12(a), more of the area of the higher curve is closer to the center; of the three curves, it would have the smallest standard deviation. The normal curve with mean 0 and standard deviation 1 is called the *standard normal* curve. The reason that this particular curve is given special attention will become obvious shortly.

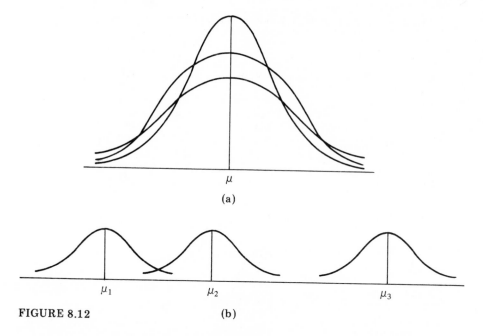

μ

(a)

μ_1 μ_2 μ_3

FIGURE 8.12 (b)

Suppose now that z denotes a quantity having a standard normal distribution with mean 0 and standard deviation 1. To calculate the probability of z taking a value between 0 and, say, 1.3, it would be necessary to determine the area indicated in Figure 8.13. Areas such as this have

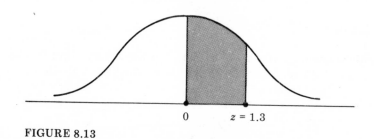

0 $z = 1.3$

FIGURE 8.13

been calculated for various values of z, some of which are given in Table 8.13. The desired probability is read from Table 8.13 for $z = 1.3$ as 0.403.

When using Table 8.13, it is essential to keep in mind that area given is *always the area under the curve from 0 to z*, as indicated in Figure 8.14.

Note that only positive values of z are given in Table 8.13. However, the probabilities for negative values of z can also be determined from the table because the curve is symmetric about 0. For example, the area under

TABLE 8.13   Areas under a Standard Normal Curve

z	Area	z	Area	z	Area
0.1	0.039	1.1	0.364	2.1	0.482
0.2	0.079	1.2	0.385	2.2	0.486
0.3	0.118	1.3	0.403	2.3	0.489
0.4	0.155	1.4	0.419	2.4	0.492
0.5	0.192	1.5	0.433	2.5	0.493
0.6	0.226	1.6	0.445	2.6	0.495
0.7	0.258	1.64	0.450	2.7	0.496
0.8	0.288	1.7	0.455	2.8	0.497
0.9	0.316	1.8	0.464	2.9	0.498
1.0	0.341	1.9	0.471	3.0	0.499
		1.96	0.475		
		2.0	0.477		

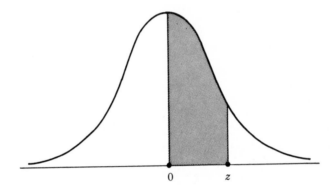

FIGURE 8.14

the standard normal curve from -2.1 to 0 is the same as that from 0 to 2.1 (Fig. 8.15). From Table 8.13, then, the probability of z taking a value between -2.1 and 0 is 0.482.

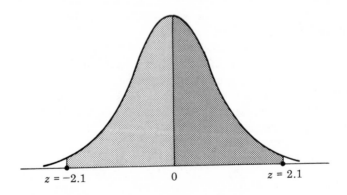

FIGURE 8.15

Other uses of Table 8.13 in the determination of probabilities are illustrated in the following examples.

EXAMPLE 1.

If z has the standard normal distribution, what is the probability that z takes a value between 0.8 and 1.5 (Fig. 8.16)?

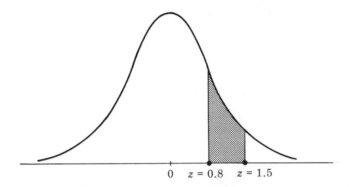

$$0 \quad z = 0.8 \quad z = 1.5$$

FIGURE 8.16

We have to determine the area indicated in Figure 8.16. From Table 8.13, the area from 0 to 1.5 is 0.433 and the area from 0 to 0.8 is 0.288. The area from 0.8 to 1.5 is therefore $0.433 - 0.288$, or 0.145.

EXAMPLE 2.

If z has the standard normal distribution, what is the probability that z takes a value between -1.3 and 2.1 (Fig. 8.17)?

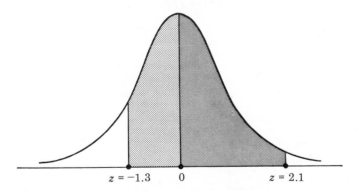

$$z = -1.3 \quad 0 \quad z = 2.1$$

FIGURE 8.17

Again, we are interested in determining the area indicated in the figure. From Table 8.13, the area between 0 and 2.1 is 0.482 and the area between -1.3 and 0 is 0.403. The desired probability is $0.482 + 0.403 = 0.885$.

EXAMPLE 3.

If z has the standard normal distribution, what is the probability that z takes a value less than -0.7 (Fig. 8.18)?

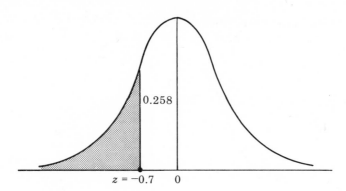

0.258

$z = -0.7$ 0

FIGURE 8.18

The total area under the curve to the left of 0 is 0.5; from Table 8.13, the area between -0.7 and 0 is 0.258. The area to the left of -0.7 is therefore $0.5 - 0.258$, or 0.242.

EXAMPLE 4.

If z has the standard normal distribution, find the value of z such that $2\frac{1}{2}\%$ of the distribution lies to the left of that value (Fig. 8.19).

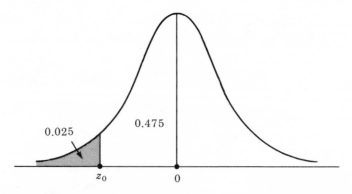

0.475

0.025

z_0 0

FIGURE 8.19

If z_0 denotes the desired value, then the area to the left of z_0 is 0.025 and the area between z_0 and 0 is 0.475. The area 0.475 in Table 8.13 corresponds to $z = 1.96$; that is, $z_0 = -1.96$.

EXERCISES

Sketching a rough "picture" for each of these exercises will probably be helpful.

1. If z has the standard normal distribution, find the probability that z takes a value

   (a)† between 0 and 1.2.   (b) between -0.6 and 0.
   (c)† between 1.5 and 2.6.   (d) between -2.4 and -0.5.
   (e)† between -1 and 2.   (f) greater than 1.8.
   (g)† less than 1.5.   (h) less than -0.8.
   (i) † greater than 2 or less than -2.

2. If z has the standard normal distribution, find the value of z for which

   (a)† $2\frac{1}{2}\%$ of the distribution lies to the right of that value.
   (b)† 5% of the distribution lies to the left of that value.
   (c) 5% of the distribution lies to the right of that value.
   (d) 10% of the distribution lies to the left of that value.

Now let x be a quantity having an arbitrary normal distribution with mean μ and standard deviation σ. If z is a quantity such that

$$z = \frac{x - \mu}{\sigma},$$

then z has the standard normal distribution. Furthermore, the area under the standard normal curve between 0 and z is the same as the area between μ and x under the normal curve for x(Fig. 8.20). Because of this relationship, Table 8.13 for the standard normal can be used to determine probabilities for arbitrary normal distributions. For example, suppose x has a

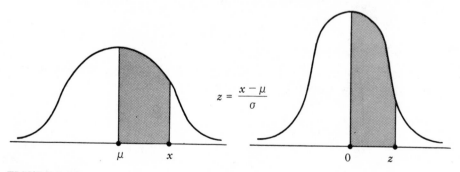

FIGURE 8.20

normal distribution with mean 20 and standard deviation 2. To determine the probability of x taking a value between 20 (the mean) and 23, we first find the z value for x = 23:

$$z = \frac{23 - 20}{2} = 1.5.$$

From Table 8.13, we find the probability to be 0.433.

EXAMPLE 5.

If x has a normal distribution with mean 16 and standard deviation 0.5, what is the probability that x takes a value less than 14.8 (Fig. 8.21)?

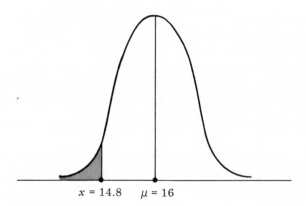

$x = 14.8$ $\mu = 16$

FIGURE 8.21

The z value for x = 14.8 is

$$z = \frac{14.8 - 16}{0.5} = \frac{-1.2}{\frac{1}{2}} = -2.4.$$

Because the area of interest is to the left of 14.8, the probability is $0.5 - 0.492 = 0.008$.

EXAMPLE 6.

If x has a normal distribution with mean 100 and standard deviation 10, what is the probability that x takes a value between 85 and 105 (Fig. 8.22)?

The corresponding z values are:

$$z = \frac{105 - 100}{10} = 0.5$$

and

$$z = \frac{85 - 100}{10} = -1.5.$$

The probability is therefore 0.433 + 0.192 = 0.625.

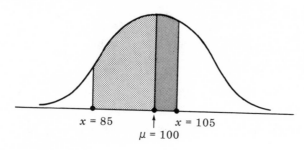

$x = 85$ $x = 105$
$\mu = 100$

FIGURE 8.22

EXAMPLE 7.

If the scores on an IQ test are normally distributed with mean 100 and standard deviation 15, what is the probability that a person chosen at random will score higher than 120?

The z value for $x = 120$ is

$$z = \frac{120 - 100}{15} = 1\tfrac{1}{3},$$

or approximately 1.3. The probability is therefore approximately 0.5 − 0.403 = 0.097.

EXAMPLE 8.

The quantity x has a normal distribution with mean 50 and standard deviation 5. Find the value of x such that 5% of the distribution is to left of that value (Fig. 8.23).

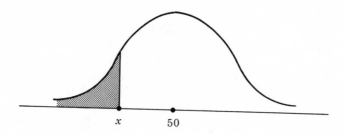

x 50

FIGURE 8.23

From Table 8.13, the value of z for which 5% of the distribution lies to the left is -1.64. The corresponding value of x satisfies the equation

$$-1.64 = \frac{x - 50}{5}.$$

Solving this equation for x gives $x = 41.8$.

EXERCISES

Sketching a rough "picture" for each of these exercises will probably be helpful.

1.† If x has a normal distribution with mean 50 and standard deviation 5, find the probability that x takes a value

    (a)   between 55 and 65.   (b)   between 47 and 50.
    (c)   between 43 and 60.   (d)   greater than 65.
    (e)   less than 52.

2. If x has a normal distribution with mean 2.3 and standard deviation of 0.5, find the probability that x takes a value

    (a)   between 2.2 and 2.4.   (b)   between 1 and 1.5.
    (c)   less than 2.   (d)   greater than 2.2.

3.† If the height of first-grade boys is normally distributed with a mean of 40 inches and standard deviation of 2 inches, what is the probability that a first-grade boy chosen at random is between 39 and 41 inches tall?

4. The amount of orange juice put into containers by a filling machine follows a normal distribution with a mean of 12 ounces and a standard deviation of $\frac{1}{3}$ ounce. What percentage of the time will the machine overflow the 13-ounce containers?

5.† (a)   Close examination over a number of years has revealed that a precision manufacturing machine produces bolts with a mean diameter of 0.25 inches and standard deviation of 0.0001 inches. What is the probability that a bolt chosen at random from the machine's output has a diameter greater than 0.2502 inches?

    (b)   The quality control inspector occasionally measures one of the bolts produced by the machine. If the diameter lies outside the range 0.2498 to 0.2502 inches, the machine is adjusted. What is the probability that the machine is adjusted at inspection time?

6. If x has a normal distribution with mean 60 and standard deviation 7, find the value of x such that

 (a)† $2\frac{1}{2}$% of the distribution lies to the right of that value.

 (b) $2\frac{1}{2}$% of the distribution lies to the left of that value.

 (c)† 10% of the distribution lies to the right of that value.

 (d) 5% of the distribution lies to the left of that value.

We conclude this section by examining one use of the normal curve in inferential statistics. The normal curve involved represents the distribution of sample means. Suppose, for example, that you are interested in the annual cost of attending college. To get more information on the topic, you send questionnaires to students on every campus in the country. When (and if!) the questionnaires return, you divide them into piles of 20 each. For each pile of 20, you calculate mean cost. These means will obviously not be the same for each possible group of 20 students, but rather will follow some sort of distribution of their own—a distribution of sample means.

In this illustration, the cost to every student in the country is the population; the costs to the 20 students in the various piles are the samples. The *size* of each sample in this case is 20.

We are interested in the distribution of the sample means. A theorem of statistics states: *If samples of size n are randomly chosen, and if the population has mean μ and standard deviation σ', then the sample means are approximately normally distributed with mean μ and standard deviation $\sigma = \sigma'/\sqrt{n}$.*

On the basis of this theorem, a test has been devised whether to accept the claim that a population has a certain mean. (For example, the claim might be that the mean annual cost of attending college in the United States is $3,500.) The test takes advantage of the fact, stated in the theorem, that the sample means and the population distributions have the same mean. A sample is taken and on the basis of its mean, a decision is made whether to accept or reject the claimed population mean.

The test begins with the assumption that the mean of the population is, in fact, the claimed mean μ. Then, the normal distribution of sample means will be as shown in Figure 8.24. The probability of a sample means taking a value in the shaded regions is 0.05.

A random sample of size n is taken and the mean $\bar{x}$ of the sample is calculated. If the value of $\bar{x}$ lies in the shaded region (the "rejection region"), the claim that the population mean is $μ$ is rejected. The reason for rejecting the claim is that the probability of obtaining a sample mean value in the rejection region is so small that the sample means must have a different distribution. The assumption on which this distribution is founded, that the population mean is $μ$, must therefore be rejected.

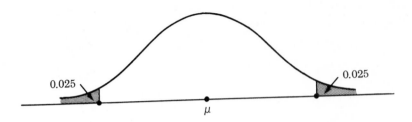

FIGURE 8.24

Before examining the use of the test in several examples, consider the standard normal corresponding to the above sample means distribution (Fig. 8.25). Here $z = (\bar{x} - \mu)/\sigma$ and $\sigma = \sigma'/\sqrt{n}$, where μ is the claimed population mean, σ' is the population standard deviation, and n is the sample size. Note that the z values for the rejection region are always $z > 1.96$ and $z < -1.96$. Consequently, one need only calculate the appropriate z values and determine whether or not they are in this rejection region.

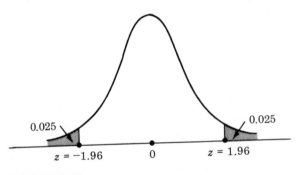

FIGURE 8.25

EXAMPLE 9.

The manufacturer of a particular type of TV picture tube claims that the mean life of its picture tubes is 7 years. A consumer group randomly picks 36 TV sets that use this tube. The mean life of these 36 picture tubes was 6 years and the standard deviation was 6 months. On this basis, should the claim of the manufacturer be accepted or rejected?

Nothing is known about σ', the standard deviation of the population, which would be the life span of all picture tubes of this type. The standard deviation of the sample (6 months) is used as an estimate for σ'. The resulting procedure is valid as long as the sample size is 30 or more.

Calculation of the z value gives

$$z = \frac{\bar{x} - \mu}{\sigma'/\sqrt{n}} = \frac{6 - 7}{0.5/\sqrt{36}} = -12.$$

Because $-12 < -1.96$, the 7-year mean must be rejected.

Note that all measurements must be in terms of the same units—months, years, and so forth. Years were used in this example.

EXAMPLE 10.

If the average age of people actually voting in a particular state is 32.5, then the most effective political campaigner would probably cater to the people in their mid-twenties to mid-thirties. A random sample of 100 actual voters has a mean age of 32 and standard deviation of 3 years. Should the assumption of a mean age of 32.5 be accepted or rejected?

The z value is

$$z = \frac{32 - 32.5}{3/\sqrt{100}} = -\tfrac{5}{3},$$

or approximately -1.67. Because $-1.96 < -1.67 < 1.96$, the mean age of 32.5 is accepted.

The test does not prove that the mean age is exactly 32.5. It simply indicates that the sample data is compatible with a mean age of 32.5. The true mean is probably not exactly 32.5; but for practical purposes, whether the true mean is 32.5 or close to 32.5 is not of great importance.

EXAMPLE 11.

A certain drug was developed to slow down one's heartbeat. The manufacturer claims that ten minutes after the drug is administered, the patient's heartbeat will be slowed by 12 beats per minute on the average.

The drug was administered at various times to 49 patients. Their heartbeats had slowed down ten minutes after receiving the drug by an average (mean) of 12.4 beats with a standard deviation of 2 beats. On this basis, should the manufacturer's claim be accepted or rejected?

The z value is

$$z = \frac{12.4 - 12}{2/\sqrt{49}} = 1.4.$$

Because $-1.96 < 1.4 < 1.96$, the manufacturer's claim is accepted. The claim is accepted in the same sense as for Example 10; the claimed mean is compatible with the sample data.

As with the sign test, the size 0.05 of the rejection region was chosen arbitrarily here and denotes the probability of rejecting the claimed mean when it is correct. All of the comments at the end of Section 8.4 on the sign test regarding the size of the rejection region and the possibility of the two types of errors—even though the decision is statistically correct—are applicable here as well; but now the assumption to be accepted or rejected is that the claimed mean is the true mean.

EXERCISES

1.† Test the claim that a family of four spends an average of $500 on its vacation, if the mean spent by a random sample of 36 such families is $550 and the standard deviation is $150.

2. A company receives a shipment of bolts that are supposed to be 0.25 inches in diameter. The receiving department picks a random sample of 49 bolts for inspection. The sample has a mean of 0.255 inches and a standard deviation of 0.01 inches.

   The bolts are not required to be precisely 0.25 inches in diameter, only that they be reasonably close. Should the company keep the shipment?

3.† The mean cost of a random sample of 36 mathematics textbooks is $13.50 and the standard deviation is $2.00. Use this information to test the claim that the average cost of mathematics textbooks is $14.00.

4. The lengths of 100 telephone calls placed through a university switchboard were recorded. The calls were randomly chosen, had a mean length of 3 minutes and had a standard deviation of 15 seconds. Test the claim that the average length of telephone calls placed through the switchboard is 4 minutes.

8.6 BEWARE OF STATISTICS!

Statistics are quoted to substantiate claims in a great variety of situations in business, academia, and government. Almost daily in newspapers and magazines and on television the public is subjected to various statistical analyses so that individuals will vote a certain way, buy a certain product, contribute to a certain cause, or act in some other desired manner. This, of course, is a proper use of statistics, but the statistics ought to be presented in an honest way.

This section indicates several methods in which statistics can be presented from a biased point of view so as to motivate the persons for whom they are intended. These methods are divided into several categories and are presented mostly by illustrations.

Unspecified Averages

Imagine yourself at a meeting of a company representative and an employee representative who are negotiating a wage agreement. The company representative maintains, "The average yearly salary of employees in this company is $12,550." The employee representative counters, "The average yearly salary of employees in this company is $8,500." Who is lying?

Actually, both could be telling the truth, but not quite the whole truth. The company representative may be presenting the mean yearly salary and the employee representative the median. The difference in the two is accounted for by the large salaries of the company executives, which cause a substantial increase in the mean.

Differences in Graphic Effect

The two graphs in Figure 8.26 present the same information concerning the amount of sales of a particular company during a twelve-month period.

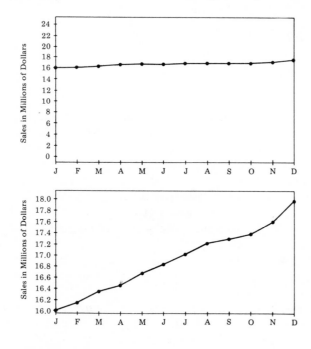

FIGURE 8.26

The difference in scales on the left side of the graph results in different visual effects from the two graphs. The first is less impressive and might be presented to the salespeople of the company by the sales manager to motivate them toward producing more sales.

The second, being more impressive, might be released to the media by the company's public relations manager so as to impress the public with the vitality of the company.

A difference of visual effects can also be obtained in the use of bar graphs in Figure 8.27 indicates. Note that each graph again presents the same information.

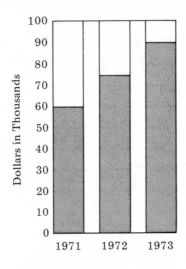 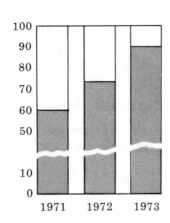

FIGURE 8.27

If you are a department manager for a particular company and these graphs represent the expenses of your department for the various years, which of the two graphs would you present to the company president to substantiate your claim that you are keeping operating expenses under reasonable control? Which would you present to the budget director to substantiate your need for an increase in operating capital for the following year, due to inflation and increased activity in your department?

A final illustration of how the visual effect of graphs can be misinterpreted is in the use of picture graphs. According to the scales on the left, each of the three graphs in Figure 8.28 give the same information. However, if the scales are ignored, and only the pictures are considered, erroneous comparisons can result.

When the dimensions of a two-dimensional figure are doubled, the area of the figure is increased fourfold. For three-dimensional figures, when the dimensions are doubled, the volume is increased eightfold. Consequently, the area of the second house in the middle graph is four times

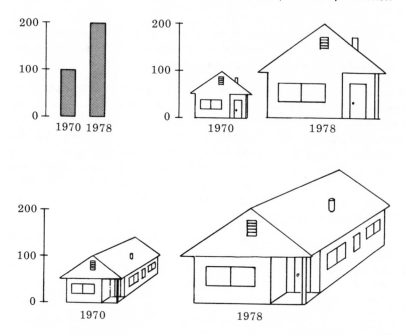

FIGURE 8.28   New Houses Completed by the ABC Construction Company

that of the first house. In the third graph, the volume of the second house is eight times that of the first house. Neither the area nor the volume interpretations convey the true fact that houses completed doubled from 1970 to 1978.

Unexplained Sample

The claim is presented:

"Seven out of ten housewives interviewed use Product X."

It is not likely that anyone would conclude that every housewife in the country was interviewed. This type of claim is the result of interviewing a smaller group of women—a sample which supposedly represents the housewives of the country.

Some legitimate questions to ask about this claim are the following:

1.   How many housewives were interviewed? 10, 50, 500, 1000, . . . ?
2.   How were the housewives chosen? At random or from additional information on how they would likely respond?
3.   If the sample was chosen at random, are the responses of all those chosen reflected in the claim, or only those responses that result in an impressive claim?

4. If the sample was chosen at random, does the claim represent the result of the first sample, or were several samples chosen until an impressive result was obtained?

None of these illustrations involve a lie in the presentation of the statistics, but either the whole truth is not told or else the statistics are presented in the manner that is most impressive for the purpose at hand. Even the statistical tests described in this chapter can be manipulated to obtain the desired results. Should one then ignore statistical claims? No, but a cautious skepticism is warranted when statistics are quoted to prove a point!

For the reader who wishes to pursue this topic further, two interesting and informative books available on the subject are:

Stephen K. Campbell, *Flaws and Fallacies in Statistical Thinking* (Englewood Cliffs, N.J.: Prentice-Hall, Inc., 1974).

Darrell Huff, *How to Lie with Statistics* (New York: W. W. Norton and Company, Inc., 1965).

8.7 SUMMARY

Statistics, in general, is concerned with collecting, describing, and interpreting data. The only aspect concerning the collection of data considered was the organization of the data into a convenient form—either in a frequency table or histogram.

The two major concepts of descriptive statistics are measures of central tendency (averages) and measures of dispersion. Those considered and their relationship are indicated in Table 8.14. The principal measures

TABLE 8.14

Measures of Central Tendency	*Measures of Dispersion*
Mean	Standard deviation
Median	Interquartile range
Mode	
Midrange	
	Range

are the mean, the standard deviation (which indicates the concentration of the data about the mean), the median, and the interquartile range (which indicates the concentration of the data about the median). The mode, midrange, and range are included in Table 8.14 mostly for the sake of completeness.

A normal curve also has a mean (μ) to indicate the middle of the curve and a standard deviation (σ) to indicate the spread of the curve about its mean. The calculation of probabilities for quantities having a normal distribution amounts to determining areas under the corresponding normal curve. Tables for the standard normal curve ($\mu = 0$, $\sigma = 1$), such as Table 8.13, give the areas for intervals from the mean 0 to positive numbers z. Because the curve is symmetric about its mean, areas for intervals from 0 to negative z are equal to areas from 0 to the corresponding positive z. Furthermore, half of the area under the curve lies to the right of the mean and half to the left; this fact is used to calculate the total area to the right or to the left of a given number z.

For an arbitrary normal curve with mean μ and standard deviation σ, the area between the mean and a given number x is the same as that under the standard normal curve between 0 and $z = (x - \mu)/\sigma$. Consequently, the tables for the standard normal can be used to calculate areas for any given normal curve.

Two aspects of inferential statistics were considered—the sign test and a test (using the normal curve) to determine whether a given number μ_0 should be accepted or rejected as a population mean. Note that the two tests follow a similar procedure:

1. An assumption is made about the population. For the sign test, this assumption is that there is no difference between the objects being compared. For the mean test, this assumption is that the population mean is the claimed mean μ_0.

2. Based on the assumption in Step 1, a rejection region is set up. The rejection region is chosen in such a way that the probability of obtaining a statistic from a sample with a value in the rejection region is quite small. The statistic in the sign test is the greater of the number of plus signs or minus signs. In the mean test, it is the mean of the sample.

3. If the statistic from the sample then takes a value in the rejection region, the original assumption is rejected. The reason for this rejection is as follows: the probability of obtaining a value in the rejection region is so small that, if the value obtained does lie in that region, something must be wrong—that is, the underlying assumption that determined the region must be invalid.

The tests are not infallible. There is always the possibility of rejecting a valid assumption and of accepting a false assumption.

Finally, a word of caution. Be skeptical of statistics that are used to motivate you or to persuade you to do something.

*"The Theory of Groups is a branch of
mathematics in which one does something
to something and then compares the result
with the result obtained from doing the
same thing to something else, or something
else to the same thing."*
        JAMES R. NEWMAN (1907–1966)

Quartz Crystals. Groups are used by physicists to develop a physical theory of crystals.

chapter 9

groups

The beginning of the theory of groups goes back to the Middle Ages, but the main impetus to the study occurred around the beginning of the 19th century, primarily due to the French mathematician Évariste Galois (Fig. 9.1). In fact, it was he who in 1830 gave the title *group* to the system. At the age of 20 Galois was shot in a duel and subsequently died. The night before the dual he frantically wrote down some of his major contributions to the subject.

The study of the group system arose from the realization that many numerical and algebraic systems possess the same structure, that is, the same relationships exist between their members with respect to some operation on their members. The group axioms capture the essential features of this structure and the development of the group system is therefore an abstract

Évariste Galois (1811–1832)

FIGURE 9.1

and axiomatic study of structure. Referring to this development, J. R. Newman writes, "It is concerned only with the fine filagree of underlying relationships; it is the most powerful instrument yet invented for illuminating structure."[1]

Applications of groups are found in a variety of areas—for example, music, campanology (bell-ringing), quantum mechanics, crystallography (the study of crystals), computer science, and coding theory. Some examples of the application of groups will be given in Section 9.4.

9.1 WHAT IS A GROUP?

The integers are made up of the counting numbers, 1, 2, 3, . . . , the negatives of the counting numbers, −1, −2, −3, . . . , and the number 0. They are denoted collectively by the set {. . . , −3, −2, −1, 0, 1, 2, 3, . . .}.

We single out four basic properties of addition of integers. (The reason for our interest in these four will become apparent shortly.)

1. The sum of two integers is another integer. For example, the sum of −5 and 8 is the integer 3.
2. When three integers are added, the grouping for the purpose of addition is immaterial. For example, $(-3 + 7) + 2 = -3 + (7 + 2)$; the same result is obtained if the sum of −3 and 7 is added to 2 as is obtained if −3 is added to the sum of 7 and 2.
3. The integer 0 plays the singular role in addition that the sum of 0 and any other integer is always the other integer. For example, $7 + 0 = 7$ and $0 + 7 = 7$.
4. The sum of any integer and its negative is always 0. For example, $9 + (-9) = 0$ and $-9 + 9 = 0$. Furthermore, the negative of any integer is also an integer.

If the operation is changed and/or the set of numbers is changed, the four corresponding properties may no longer be present. For example, suppose that the operation is changed to subtraction, but the set is still the integers. Then the corresponding Property 2 no longer holds. For instance,

$$3 - (4 - 5) \neq (3 - 4) - 5.$$

The left side is equal to 4; the right is equal to −6.

Or suppose that the operation is again addition but the set of numbers is taken to be natural numbers, {1, 2, 3, 4, 5, . . .}. The corresponding Property 3 no longer holds as 0 is not a natural number. What's more, there is no natural number that behaves in the way required for Property 3.

But if the operation is multiplication and the set of numbers is chosen to be the rational number (that is, the numbers which can be written

---

[1] James R. Newman, *The World of Mathematics* (New York: Simon and Schuster, Inc., 1956), vol. 3, p. 1534.

as the quotient of two integers such as $\frac{2}{3}, -\frac{3}{4}$, or $2 = \frac{2}{1}$), with 0 removed, then the four corresponding properties do hold.

1. The product of two rational numbers is another rational number. For example, $\frac{2}{3} \times (-\frac{3}{4}) = -\frac{6}{12} = -\frac{1}{2}$, and $-\frac{1}{2}$ is a rational number.
2. When three rational numbers are multiplied, the grouping for the purpose of multiplication is immaterial. For example, $(\frac{2}{3} \times \frac{5}{7}) \times \frac{1}{4} = \frac{2}{3} \times (\frac{5}{7} \times \frac{1}{4})$.
3. The product of the rational number 1 with any other rational number is always the other rational number. For example, $1 \times (-\frac{2}{3}) = -\frac{2}{3} \times 1 = -\frac{2}{3}$.
4. The product of any rational number (except for 0) with its reciprocal is always 1. For example, $\frac{2}{3} \times \frac{3}{2} = 1$ and $\frac{3}{2} \times \frac{2}{3} = 1$. (*Note:* 0 has no reciprocal; this is the reason it was removed from consideration.)

The rational numbers (with 0 removed) together with multiplication are therefore considered to have the same mathematical structure as do the integers with addition, in the sense that the above four corresponding properties are present in each case. We saw that neither the integers with subtraction nor the natural numbers with addition share this same structure. However, a surprising variety of mathematical creations (some of which we will examine later) do. Group theory is the study of this common structure.

Basically, a group consists of a *set G* with an *operation* on pairs of members of *G* that satisfies axioms similar to the four properties considered above. The idea of an operation is not new to anyone who has studied arithmetic; addition, subtraction, multiplication, and division are all examples of operations on pairs of numbers. It is also apparent that the order in which the numbers are taken will often affect the result: 2 minus 3 does not give the same result as 3 minus 2. To indicate that the operations are to be performed in a particular order, $2 \div 3$ rather than $3 \div 2$, the operation will be said to be on *ordered* pairs. The ordered pair (2,3) is not the same as the ordered pair (3,2); subtraction on the ordered pair (2,3) leads to the result of -1, whereas subtraction on the ordered pair (3,2) leads to the result of 1.

Formally, a *group* is a set *G* together with an operation $*$ on all ordered pairs of elements of *G* that satisfy the axioms given on the following page. The members of *G* will be denoted by lower case letters, *a, b, c, d, e, f, ...* The result of the operation $*$ on any ordered pair (a,b) of members of *G* will be denoted by $a * b$ (just as the quotient 2 divided by 3 is denoted by $2 \div 3$).

Axiom 1.   For any two members, a and b, of G, $a * b$ is also a member of G.

Axiom 2.   For any three members, a, b, c, of G, $(a * b) * c = a * (b * c)$.

Axiom 3.   There is an element e in G (called the identity) such that for each element a in G, $a * e = e * a = a$.

Axiom 4.   For each element a in G there is an element a^{-1} in G (called the *inverse* of a) such that $a * a^{-1} = a^{-1} * a = e$.

Compare the four axioms with the four properties of addition of integers. Each axiom states the corresponding property in a more general form, with "$*$" in place of "$+$," "e" in place of "0," and "a^{-1}" in place of "$-a$." Similar replacements can be made for multiplication of the rational numbers, with 0 removed.

The set of integers $G = \{\ldots, -3, -2, -1, 0, 1, 2, 3, \ldots\}$, together with the operation addition ($+$), forms a group—as would be expected. The identity element is 0 and the inverse of an integer a is $-a$.

The integers with division as the operation would not form a group because the quotient of two integers need not be another integer (Axiom 1); for example, $2 \div 7$ would not be in the set G.

The name of the system, *group*, is suggested by Axiom 1, which requires that the result of the operation $*$ on two members of the group be in the group—that is, the set G possesses a certain cohesiveness with respect to the operation $*$.

Up to now, we have thought of a set G as a set of numbers and the operation $*$ to be one of the usual arithmetic operations, addition, subtraction, multiplication, or division. These restrictions are not required by the definition of a group. In the next section, we will see that the elements of set G and the operation $*$ can be quite different—surprisingly different in some cases.

EXERCISES

1.† The set G of rational numbers, with 0 removed, together with the operation multiplication form a group. What is the meaning of "$*$," "e," and "a^{-1}" in the group axioms in this case?

2. Why are the following *not* examples of groups?

(a) The set of natural numbers, $\{1, 2, 3, 4, 5, \ldots\}$, with operation addition.

(b)† The set of odd integers, $\{\ldots, -3, -1, 1, 3, 5, \ldots\}$, with operation multiplication. With operation addition.

(c)   The set of rational numbers (that is, numbers that can be written as the quotient of two integers) greater than or equal to zero, with operation multiplication.

*3.† In order that a given set G and operation $*$ do not form a group, it is sufficient that only one of the axioms does not hold. Verify that if G is the set of rational numbers and $*$ is division, none of the axioms hold.

9.2 EXAMPLES OF GROUPS

EXAMPLE 1.

The set $G = \{a, b, c, d\}$ with operation defined by the table in Figure 9.2 constitutes a group. The use of the table is illustrated in

*	a	b	c	d
a	a	b	c	d
b	b	d	a	c
c	c	a	d	b
d	d	c	b	a

FIGURE 9.2

Figure 9.3 by determining the value of $b * c$. Find the row which contains b on the left; find the column which contains c on the top. This row and column intersect at a; hence, $b * c = a$.

*	a	b	c	d
a	a	b	c	d
b	b	d	a	c
c	c	a	d	b
d	d	c	b	a

FIGURE 9.3

The identity in this example is a; the inverse of c, for example, would be b because $b * c = c * b = a$.

Note that the column under the identity, a, is the same as the column at the left of the table, and that the row following a is the same as the row at the very top of the table. Because of the nature of the identity, this match-up of columns and rows will occur in every group when the operation is given in tabular form and makes the identity easy to identify.

EXAMPLE 2.

Suppose that we have a triangle with three sides of equal length, and with the corners numbered for identification purposes as shown in Figure 9.4. A rotation through 180° about the axis from

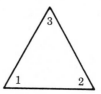

FIGURE 9.4

vertex 3 to the opposite side, as indicated in Figure 9.5, results in a triangle that looks exactly like the original except the location of the vertices has changed.

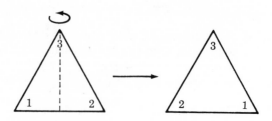

FIGURE 9.5

Similarly, if the original triangle is rotated counterclockwise about its center through 120 degrees, the result would be as shown in Figure 9.6.

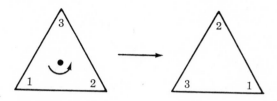

FIGURE 9.6

Now perform these two rotations in succession (Fig. 9.7).

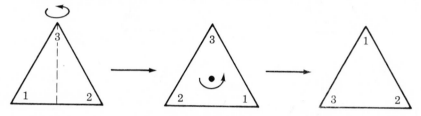

FIGURE 9.7

The result of this composition is the same as if the original triangle had been rotated about its axis through vertex 2 (Fig. 9.8).

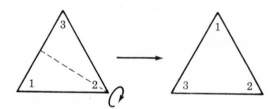

FIGURE 9.8

Each of these rotations results in a triangle that looks like the original triangle, except for the numbers of the corners. The set of all such rotations (there are six of them), together with the operation of composition illustrated above, forms a group. The six members of this group and their designations are indicated in Figure 9.9. The members of this group are called the *symmetries of a triangle.*

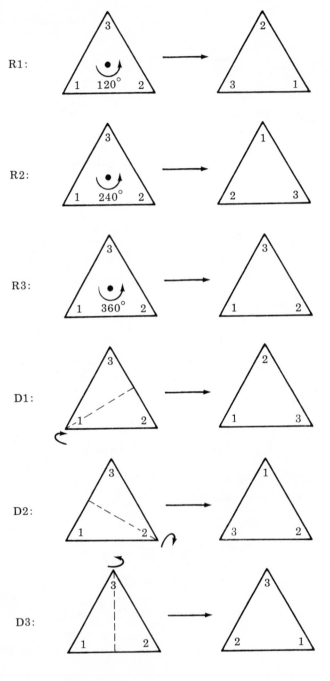

FIGURE 9.9

The result of R2 * D3 would be determined by performing D3 on the original triangle and then R2 on this result (Fig. 9.10). This is the same as would be obtained by performing D1 on the original triangle. Hence, R2 * D3 = D1.

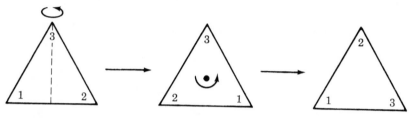

FIGURE 9.10

Similarly, D2 * R2 = D3 (Fig. 9.11).

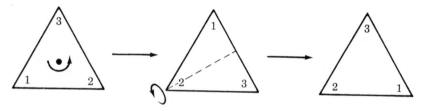

FIGURE 9.11

The result of the composition of the various symmetries of a triangle is given in Table 9.1.

TABLE 9.1

*	R1	R2	R3	D1	D2	D3
R1	R2	R3	R1	D3	D1	D2
R2	R3	R1	R2	D2	D3	D1
R3	R1	R2	R3	D1	D2	D3
D1	D3	D2	D1	R3	R2	R1
D2	D1	D3	D2	R1	R3	R2
D3	D2	D1	D3	R2	R1	R3

From Table 9.1, we see that D2 * D1 = R1, but D1 * D2 = R2.

Conclusion: In general, the order in which the group operation is performed is important!

EXAMPLE 3.

When it is 10 o'clock, the little hand of a clock points to 10, and seven hours later it points to 5 (Fig. 9.12).

FIGURE 9.12

In this sense, 10 + 7 = 5. In the same way,

$$8 + 6 = 2,$$
$$3 + 4 = 7,$$
$$5 + 20 = 1,$$
$$6 + 33 = 3,$$

and $\quad 7 + 5 = 0,$

the latter because we have chosen to replace 12 with 0 on our clock. The sum in each case is obtained by moving the little hand around the clock through the appropriate number of hours.

Because a complete circuit of the little hand around the clock corresponds to an increase of 12, we can get the same results by performing addition in the usual way and then subtracting the largest multiple of 12 that is less than the sum obtained:

$$10 + 7 = 17 \quad \text{and} \quad 17 - 12 = 5$$
$$8 + 6 = 14 \quad \text{and} \quad 14 - 12 = 2$$
$$3 + 4 = 7 \quad \text{and} \quad 7 - 0 = 7$$
$$5 + 20 = 25 \quad \text{and} \quad 25 - 24 = 1$$
$$6 + 33 = 39 \quad \text{and} \quad 39 - 36 = 3$$
$$7 + 5 = 12 \quad \text{and} \quad 12 - 12 = 0.$$

Arithmetic of this type is called *modular* arithmetic, and, in this particular case, *addition modulo 12.*

The same type of addition can be performed with 12 replaced by any other natural number. Table 9.2 gives the sums for addition modulo 5. The sum 3 + 4 modulo 5, for example, is given by

$$3 + 4 = 7 - 5 = 2 \ (\text{mod } 5).$$

The notation "(mod 5)" is included to indicate the type of addition being performed.

TABLE 9.2

+ (mod 5)	0	1	2	3	4
0	0	1	2	3	4
1	1	2	3	4	0
2	2	3	4	0	1
3	3	4	0	1	2
4	4	0	1	2	3

By now you must suspect that $G = \{0, 1, 2, 3, 4\}$, together with addition modulo 5, is a group; and such is the case. From Table 9.2, we see that the identity is 0. Can you construct a formula for determining the inverse of an element in this group?

For any natural number n, the set $G = \{0, 1, 2, 3, \ldots, n - 1\}$, together with addition modulo n, forms a group.

Modular multiplication is performed in much the same way as addition—ordinary multiplication followed by subtraction of the largest multiple. For example,

$$6 \times 4 = 24 - 20 = 4 \ (\text{mod } 5).$$

Do you think that modular multiplication enters into the formation of groups in the same way as modular addition?

EXAMPLE 4.

This final group example is one that is used in information theory or coding theory and will be considered again in Section 9.4 on applications of groups.

The operation in this example, while not addition modulo 2, does depend on the addition of 0 and 1 modulo 2. We give this addition first:

$$0 + 0 = 0 \ (\text{mod } 2)$$
$$0 + 1 = 1 \ (\text{mod } 2)$$
$$1 + 0 = 1 \ (\text{mod } 2)$$
$$1 + 1 = 0 \ (\text{mod } 2).$$

The elements of the set G are now all of the possible ordered triples of 0 and 1:

$$G = \{000, 001, 010, 100, 011, 101, 110, 111\}.$$

The group operation $*$ on any two elements is performed by modulo 2 addition of their corresponding members. For example, $101 * 011 = 110$, because:

$$1 + 0 = 1 \ (\text{mod } 2) \ \text{(adding the first members)}$$
$$0 + 1 = 1 \ (\text{mod } 2) \ \text{(adding the second members)}$$
$$1 + 1 = 0 \ (\text{mod } 2) \ \text{(adding the third members)}.$$

In the same way, $111 * 111 = 000$. Note that there are no carries, as in ordinary arithmetic or in binary arithmetic.

The identity in this case is 000. The inverse of 010, for example, is also 010 since $010 * 010 = 000$. From the last paragraph, we see that the inverse of 111 is also 111. In fact, every element in this group is its own inverse.

The elements of G are all ordered arrangements of 0 and 1, three digits at a time. The arrangements could also be two at a time, four at a time, ten at a time, and so forth. In each case, another group would result with an operation similar to the operation in this example.

Other examples of groups are given in the exercises.

EXERCISES

1. The set $G = \{a, b, c, d\}$, with operation $*$ defined by Table 9.3, is a group.

(a)† Determine $c * d$, $b * a$.

(b)† Determine $b * (a * b)$, $(c * c) * d$.

(c) What is the identity element?

(d)† What is the inverse of a? Of c?

TABLE 9.3

*	a	b	c	d
a	a	b	c	d
b	b	c	d	a
c	c	d	a	b
d	d	a	b	c

2. In the group of symmetries of a triangle (Table 9.1),

   (a)† determine R3 * D2, R2 * R3.
   (b)† determine R1 * (D2 * R2), D1 * (D2 * D3).
   (c) what is the identity element?
   (d)† what is the inverse of R1? Of D2? Of R3?

3. Construct an addition table for the group with $G = \{0, 1, 2, 3, 4, 5\}$ and addition modulo 6.

4.† $G = \{0, 1, 2, 3\}$ with multiplication modulo 4 is not a group. Verify that Axiom 4 is not satisfied using Table 9.4.

TABLE 9.4

X (mod 4)	0	1	2	3
0	0	0	0	0
1	0	1	2	3
2	0	2	0	2
3	0	3	2	1

5. $G = \{00, 01, 10, 11\}$. The result of the operation * on any two elements is obtained by modulo 2 addition of corresponding members, as in Example 4 of this section.

   (a) Construct a table for *.
   (b) What is the identity?
   (c) Determine the inverse of each element.

6. A 2 by 2 integral matrix is a rectangular array of the form

$$\begin{bmatrix} a & b \\ c & d \end{bmatrix}$$

where a, b, c, and d are integers. The sum of two such matrices is obtained by adding corresponding elements; for example,

$$\begin{bmatrix} -1 & -2 \\ 0 & 3 \end{bmatrix} + \begin{bmatrix} 2 & 2 \\ 3 & 0 \end{bmatrix} = \begin{bmatrix} -1+2 & -2+2 \\ 0+3 & 3+0 \end{bmatrix} = \begin{bmatrix} 1 & 0 \\ 3 & 3 \end{bmatrix}.$$

The set of all such 2 by 2 matrices, together with this addition operation, form a group.

(a) Determine the following sums:

i.† $\begin{bmatrix} 2 & 7 \\ 3 & -6 \end{bmatrix} + \begin{bmatrix} -1 & 0 \\ -3 & -4 \end{bmatrix}$

ii. $\begin{bmatrix} 12 & -7 \\ 10 & 4 \end{bmatrix} + \begin{bmatrix} -10 & 2 \\ 4 & -4 \end{bmatrix}$

iii.† $\left(\begin{bmatrix} 1 & 2 \\ 3 & 4 \end{bmatrix} + \begin{bmatrix} 5 & 6 \\ 7 & 8 \end{bmatrix} \right) + \begin{bmatrix} -6 & -8 \\ -10 & -12 \end{bmatrix}$

iv. $\begin{bmatrix} 1 & 2 \\ 3 & 4 \end{bmatrix} + \left(\begin{bmatrix} 5 & 6 \\ 7 & 8 \end{bmatrix} + \begin{bmatrix} -6 & -8 \\ -10 & -12 \end{bmatrix} \right)$

(b) What is the identity element in this group?

(c) What is the inverse of:

i.† $\begin{bmatrix} 2 & -2 \\ -5 & 7 \end{bmatrix}$

ii. $\begin{bmatrix} 3 & 10 \\ 11 & 17 \end{bmatrix}$

iii.† $\begin{bmatrix} 1 & 2 \\ 3 & 4 \end{bmatrix} + \begin{bmatrix} -2 & -4 \\ -1 & -3 \end{bmatrix}$

iv. $\begin{bmatrix} 1 & 0 \\ 0 & 1 \end{bmatrix} + \begin{bmatrix} 0 & 1 \\ 1 & 0 \end{bmatrix}$

7. Verify that $G = \{1, -1\}$, together with ordinary multiplication, forms a group by verifying that each axiom is satisfied. (You will have to consider eight cases to show that Axiom 2 is satisfied.)

*8.   Let $G = \{1, -1, i, -i\}$. If $i \times i = -1$ and $1 \times i = i \times 1 = i$, then G, to-
gether with the operation multiplication ($\times$), forms a group.

   (a)† Construct a table for multiplication, assuming that the usual
        properties for multiplying by negatives hold.
   (b)   What is the identity element?
   (c)   What is the inverse of i? Of $-i$? Of 1? Of -1?

*9.   Let P and Z be the fixed points indicated on the circle in Figure 9.13.

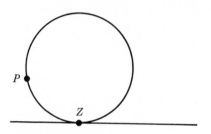

FIGURE 9.13

For any two distinct points on the circle, say A and B, $A * B$ is determined as follows: (1) draw the line through A and B, and designate
the point of intersection with the tangent line by Y, and (2) draw the
line through P and Y (Fig. 9.14).

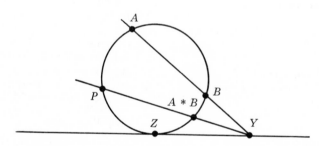

FIGURE 9.14

$A * B$ is the second point at which the line through P and Y intersects
the circle; if this line is tangent to the circle at $A * P$, then $A * B = P$.

   (a)   Verify that P is the identity.
   (b)   Locate the inverse of the point A indicated in Figure 9.14.

*10.† (For readers familiar with binary arithmetic as in Section 3.7.)
      Discuss three different meanings of the expression "1 + 1." When the
      expression "1 + 1" is encountered with no further explanation, which
      of the three meanings is presumed?

9.3 SOME THEOREMS FOR GROUPS

The easiest group to work with, because it is the most familiar, is the set of integers with the operation addition. Each of the following theorems concerning groups could arise by considering which properties hold for the integers under addition and then conjecturing that the same properties might hold for every group.

In this way, the first theorem could arise by considering that for any three integers, d, f, and g, if $d + f = g$, then we know that f must be the same as $-d + g$; for instance, if $3 + f = 7$, then $f = -3 + 7$, or $f = 4$. That a similar property holds in every group is the statement of Theorem 1. The proof of the theorem is the verification that the theorem is a consequence of the axioms and definition.

Theorem 1.   For any elements, d, f, and g, in a group G,
if $d * f = g$, then $f = d^{-1} * g$.

The proofs for the theorems will be given more formally than they were previously. They will be given one statement at a time along with an analysis of the reasons why the statement is valid. The statements are numbered for the sake of reference.

Proof	*Analysis*
1. Assume $d * f = g$.	Statement 1 assumes the conditional part of the statement of the theorem.
2. Then, $d^{-1} * (d * f) = d^{-1} * g$.	Because of the equality expressed in Statement 1, d^{-1} operating on $d * f$ will be equal to d^{-1} operating on g.
3. It follows that $(d^{-1} * d) * f = d^{-1} * g$.	Because of Axiom 2, the terms on the left side of Statement 2 may be regrouped. In terms of d^{-1}, d and f, Axiom 2 states "$d^{-1} * (d * f) = (d^{-1} * d) * f$."
4. Hence, $e * f = d^{-1} * g$.	By Axiom 4, $(d^{-1} * d)$ in the left side of Statement 3 is the same as the identity element e.
5. Consequently, $f = d^{-1} * g$.	The left side of Statement 4 is the same as f by Axiom 3.
6. Therefore, if $d * f = g$, then $f = d^{-1} * g$.	This is the statement of the theorem. By assuming the conditional part of the theorem in Statement 1, we were able to arrive at the conclusion of the theorem in Statement 5 by means of the axioms.

If d, f, and g are members of the group of integers such that $g = (-d) + f$, we know that $d + g = f$; for instance, if $7 = -4 + f$, then $4 + 7 = f$, or $f = 11$. Theorem 2 states that a similar property holds in every group.

Theorem 2.   For any elements, d, f, and g, in a group G, if $g = d^{-1} * f$, then $d * g = f$.

Proof	Analysis
1. Suppose that $g = d^{-1} * f$.	Statement 1 assumes the conditional part of the theorem.
2. Then, $d * g = d * (d^{-1} * f)$.	Because of the equality expressed in Statement 1, d operating on g will be equal to d operating on $d^{-1} * f$.
3. It follows that $d * g = (d * d^{-1}) * f$	Because of Axiom 2, the terms on the right of Statement 2 may be re-grouped.
4. Therefore, $d * g = e * f$.	By Axiom 4, $d * d^{-1}$ in Statement 3 is the same as the identity e.
5. Hence, $d * g = f$.	$e * f$ in Statement 4 is the same as f, by Axiom 3.
6. Consequently, if $g = d^{-1} * f$, then $d * g = f$.	This is the statement of the theorem. From the assumption of the conditional part in Statement 1, we obtained the conclusion in Statement 5 by means of the axioms.

In the exercises, the reader is asked to give an analysis for each statement in the proof of Theorem 3.

Theorem 3.   For any elements, d, f, and g, in a group G, if $d * f = d * g$, then $f = g$.

1. Assume that $d * f = d * g$.
2. Then, $d^{-1} * (d * f) = d^{-1} * (d * g)$.
3. Therefore, $(d^{-1} * d) * f = (d^{-1} * d) * g$.
4. Consequently, $e * f = e * g$.
5. That is, $f = g$.
6. Therefore, if $d * f = d * g$, then $f = g$.

Theorem 4.   The identity element in a group G is unique; that is, if e_1 and e_2 are identities, then $e_1 = e_2$.

Proof	Analysis
1. Assume that e_1 and e_2 are identities.	This is the assumption of the conditional part of the theorem.

Proof	*Analysis*
2. Then, $e_2 * e_1 = e_1$.	Statement 2 follows from Statement 1 by Axiom 3. e_2 operating on any member of G must result in that member. (This statement uses the fact that e_2 is an identity.)
3. Also, $e_2 * e_1 = e_2$.	Same as Statement 2. e_1 operated on by any member of G must result in that member. (This statement uses the fact that e_1 is an identity.)
4. Consequently, $e_1 = e_2$.	This follows from Statements 2 and 3. Because e_1 and e_2 are each equal to $e_2 * e_1$, they must be equal to each other.
5. Therefore, if e_1 and e_2 are identities, then $e_1 = e_2$.	This is the statement of the theorem. From the assumption of the conditional part of the theorem, we obtained the conclusion in Statement 4 by means of the axioms.

A second proof of Theorem 4 will also be given. This second proof is in the form that is usually encountered in mathematics.

Proof (of Theorem 4): Assume that e_1 and e_2 are identities in a group G. Then, $e_2 * e_1 = e_1$ and $e_2 * e_1 = e_2$, so that $e_1 = e_2$.

This second proof is simply the translation of the first proof into the form normally used. Note that it contains no mathematical expressions or equations that are not part of a sentence! In a sense, this second proof is in skeleton form, insofar as it contains a minimum amount of mathematical analysis. The concentration is on the valid mathematical statements. This is the standard procedure in presenting mathematical proofs; the reader is assumed to be able to make a thorough analysis, if such is desired.

Several other theorems valid in the system of groups will be given in the exercises. It is hoped that, from the statements of the theorems, the reader will agree that the quotation at the beginning of this chapter is appropriate!

In this system, note that *operation* is an undefined term, while *group*, *identity*, and *inverse* are defined terms. The axioms and theorems are obvious.

This is only the barest beginning of the development of group theory. Volumes have been written on the subject, as well as thousands of research articles. Our purpose here is to see how a mathematical system is developed, not its full development.

EXERCISES

1.† Give an analysis for each valid statement in the proof of Theorem 3 given in the text.

2.† Complete the proof of Theorem 5 below. Give an analysis for each statement in the proof. (Compare with Theorem 2.)

Theorem 5.   For any elements, d, f, and g, in a group G, if $g = f * d^{-1}$, then $g * d = f$.

Proof:

1. Assume that $g = f * d^{-1}$.
2. Then, $g * d = (f * d^{-1}) * d$.
3.
4.
5.
6.

3. Complete the proof of Theorem 6 below. Give an analysis for each statement in the proof. (Compare with Theorem 1.)

Theorem 6.   For any elements, d, f, and g, in a group G, if $f * d = g$, then $f = g * d^{-1}$.

Proof:

1. Assume that $f * d = g$.
2. Then, $(f * d) * d^{-1} = g * d^{-1}$.
3.
4.
5.
6.

4.† Complete the proof of Theorem 7 below. (Compare with Theorem 3.)

Theorem 7.   For any elements, d, f, and g, in a group G, if $f * d = g * d$, then $f = g$.

Proof:

1. Assume that $f * d = g * d$.
2. Then, $(f * d) * d^{-1} = (g * d) * d^{-1}$.
3.
4.

5.
6.

5. Complete the proof of Theorem 8 below.

Theorem 8.   For any element a in a group G there is only one
inverse; that is, if d and f are inverses of a, then
$d = f$.

Proof:

1. Suppose that d and f are inverses of a.
2. Then, $d * a = e$.
3. Also, $f * a = e$.
4. Therefore, $d * a = f * a$.
5.
6.
7.
8.
9.

6. Rewrite the proof of Theorem 3 in the form of the second proof
    given in this section for Theorem 4.

9.4 APPLICATIONS OF GROUPS

Most applications of groups occur in fairly technical areas. To fully appreciate an application of this type, one must be familiar both with groups
and the technical area in which the application occurs. Lacking the technical knowledge required for the two applications to be considered in this
section, we will not be able to investigate these applications in complete
detail. However, we shall be able to appreciate the groups involved and,
in a general way, how the groups are involved in solving the physical
problems.

EXAMPLE 1.

Communication between two computers has now become a fairly
common occurrence. The computer in a spacecraft communicates
with a computer in the control center. Computers in distant parts of
the world communicate with each other via orbiting satellites. Computers closer to each other communicate via electric cables.

One way in which this is done is for the information to be coded into "words" consisting of a series of 0's and 1's. The 0's and 1's are then transmitted by electric signals. Electrical disturbances in the atmosphere, in the cables, or in the computers themselves can distort these signals. The result is that the receiving computer recieves 0 signals as 1's and 1 signals as 0's. A high degree of accuracy is usually required, so the resulting faulty information cannot be tolerated.

The problem is how to decrease the likelihood that the receiving computer receives incorrect information. One solution is to increase the word length of the information sent in such a way that the receiving computer recognizes that it has received a faulty signal. For example, if the word 1011 is to be sent, each symbol would be repeated three times, and the actual transmission would be the signals for 111000111111.

If the signals received were 101000111111, the receiving computer would recognize that the first three numbers, 101, formed an improper pattern and, if the computer were so programmed, could automatically make the correction to 111 on the basis that this was the most likely transmitted pattern.

Or suppose that all of the words sent were two symbols long. The entire vocabulary would be 00, 01, 10, and 11. The transmitting computer might code these as follows.

$$00 \rightarrow 000$$
$$01 \rightarrow 011$$
$$10 \rightarrow 101$$
$$11 \rightarrow 110$$

Signals for the symbols on the right would then be transmitted.

You may have recognized that the vocabulary of the transmitting computer in this case are the elements in the group of Exercise 5, Section 9.2. The code is formed by attaching a third symbol to each word in the vocabulary. The third symbol in each case is the sum modulo 2 of the 1's in the word to be sent.

It is easy for the receiving computer to calculate this sum for the words it receives. If the sum does not correspond to the symbols it received, it can recognize and/or correct the faulty information.

You may have noticed also in our illustration that the lengthened words on the right are elements of the group in Example 4, Section 9.2. Consequently, two groups are involved in our coding arrangement, one for the vocabulary and one for the encoded vocabulary. The same is true regardless of the length of the words involved. In our illustration, words of length two and of length three were used for the sake of simplicity.

One of the major problems in coding theory is the question of which codes to select to obtain the greatest possible accuracy. The details of how groups are used to seek a solution are too complicated for us to consider here. The point is that groups enter in a natural way and form the mathematical model for the problem under consideration. All of the theory of groups is then available for seeking a solution.

The group involved in the next example is very similar to the group of symmetries of a triangle considered in Example 2, Section 9.2. In place of the rotations of a triangle, the elements in the group here are the rotations of a regular tetrahedron, which is a three-dimensional figure, the four faces of which are congruent equilateral triangles (Fig. 9.15). A clockwise

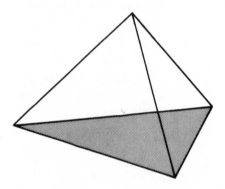

FIGURE 9.15

rotation through $120°$ along the axis indicated by the broken line in Figure 9.16 through the peak of the pyramid to the midpoint of the base

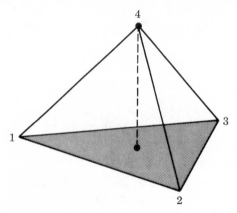

FIGURE 9.16

triangle results in Figure 9.17, which looks the same as the original figure (except for the corners, which are numbered only for identification purposes). A clockwise rotation of the original figure through 240° about

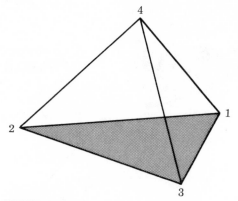

FIGURE 9.17

the same axis results in Figure 9.18. Such a rotation through 360° returns each corner to its original position.

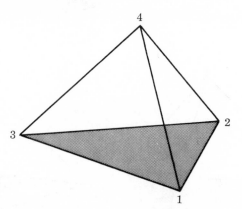

FIGURE 9.18

Corresponding rotations about the axes indicated in Figure 9.19 (each axis being between one of the vertices and the midpoint of the opposite side) give similar results.

In addition, a clockwise rotation through 180° along the axis from the midpoint of the back edge of the base triangle through the midpoint of the opposite edge (Fig. 9.20) results in Figure 9.21.

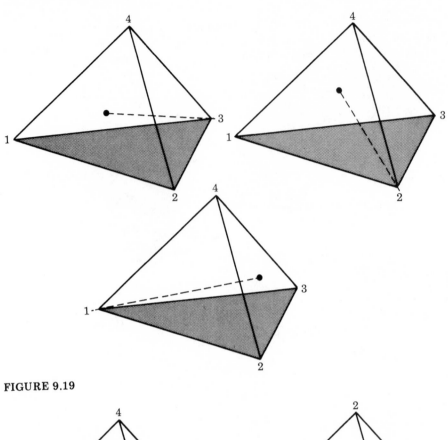

FIGURE 9.19

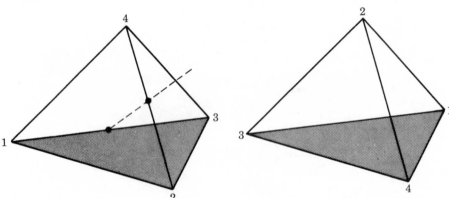

FIGURE 9.20   FIGURE 9.21

Again, corresponding rotations about the axes indicated in Figure 9.22 give similar results.

Considering any two rotations to be the same if they result in the same location of the respective corners, there are a total of 12 distinct rotations. Together with an operation of composition similar to that for the rotations of a triangle, this set of 12 rotations forms a group.

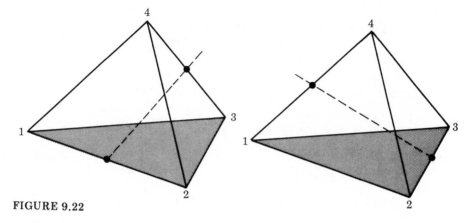

FIGURE 9.22

EXAMPLE 2.

The structure of a molecule of methane (CH$_4$) determines a regular tetrahedron, with an atom of hydrogen (H) at each vertex and an atom of carbon (C) in the center of the array of hydrogen atoms (Fig. 9.23)

FIGURE 9.23

When the methane is in a state of radiation, the molecules vibrate (as if the connections between the atoms were replaced by springs) because of forces being applied to the various atoms. In physics, the possible modes (types) of vibration are studied for the purpose of analyzing the spectral properties of methane.

The analysis of these modes of vibration is facilitated by considering the rotations of a molecule of methane in the manner just described for a regular tetrahedron. Hence, the group becomes the mathematical model of these rotations, and the complete theory of groups is therefore available for this analysis.

The full description of this application is, of course, beyond the scope of our presentation here. The examples considered here are intended to illustrate how far-reaching the applications of groups have become.

9.5 SUMMARY

A group is a set G, together with an operation $*$ on ordered pairs of elements of G, satisfying the following four axioms:

Axiom 1.   For any two members, a and b, of G, $a * b$ is also a member of G.

Axiom 2.   For any three members, a, b, c, of G, $(a * b) * c = a * (b * c)$.

Axiom 3.   There is an identity element e in G such that $a * e = e * a = a$ for each element a in G.

Axiom 4.   For each element a in G, there is an inverse of a (denoted by a^{-1}) in G such that $a * a^{-1} = a^{-1} * a = e$.

As indicated in the examples and exercises of this chapter, there is quite a variety in the types of groups. However, the most familiar is the group consisting of the set of integers with the operation addition. In this case, the identity e is 0, the inverse of an integer a is $-a$, and the operation $*$ is $+$.

The theorems, as the consequences of the axioms, examine further the group structure required by the axioms. Those theorems considered in this chapter are suggested by the corresponding properties of the group of integers with operation addition.

Applications of groups are generally not elementary. We considered only a type of group used in coding theory and a group used in the study of properties of methane.

"I know what you're thinking about," said Tweedledum; "but it isn't so, nohow." "Contrariwise," continued Tweedledee, "if it was so, it might be; and if it were so, it would be; but as it isn't, it ain't. That's logic."

Lewis Carroll, *Through the Looking-Glass*

chapter 10

logic

How do proofs verify that theorems are logical consequences of the axioms? By what set of rules does one determine whether one set of statements constitutes a proof whereas another set of statements does not? The answer is, of course, the rules of logic. An understanding of logic is essential for a full appreciation of the manner in which a mathematical system is developed and of the rigor that is usually associated with mathematics. It is because of this basic importance that this chapter is devoted to the study of logic, the study of the process of correct reasoning.

The process of reasoning from the axioms to their consequences is an exercise in *deductive* logic. Consequently, it is deductive logic with which we shall be concerned here.

The formalization of the rules of logic was first undertaken by Aristotle (around 350 b.c.) and is still being studied and enlarged upon today by philosophers and mathematicians alike. Although our interest in these rules is primarily from the point of view of their role in mathematics, hopefully our study of them will enable us to recognize and use convincing argumentation in other situations as well.

10.1 VENN DIAGRAMS AND THE PROCESS OF REASONING

An introductory approach to valid reasoning, using Venn diagrams, is illustrated by the next several examples.

EXAMPLE 1.

Consider the two premises (assumptions):

All students study mathematics.
All economics majors are students.

If M denotes the set of all people who study mathematics, S the set of all students, and E the set of all economics majors, the relationships between these sets stated by the premises are represented in Figure 10.1. The points inside the circle M represent all people who

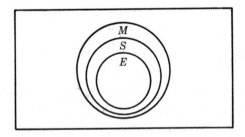

FIGURE 10.1

study mathematics; the points inside the circle S represent all students, and the points inside the circle E represent all economics majors. The points inside the rectangle represent all people—the universal set.

Is the statement "All economics majors study mathematics" a valid conclusion? In Figure 10.1, note that the assumed relationships between the sets require that all points representing economics majors be contained in the set M; that is, that all economics majors do indeed study mathematics.

In the same way, Figure 10.1 indicates that "All people who study mathematics are economics majors" is not a valid conclusion, because the assumptions do not require that each point of M be a point of E.

In such a scheme, individual members of a set are represented by points, as illustrated in the next example.

EXAMPLE 2.

The premises are:

All men are workers.
Patrick is a man.

The diagrammatic representation is shown in Figure 10.2. In this case, "Patrick is a worker" is a valid conclusion because the relationships assumed by the premises require that the point P be contained in the set of workers.

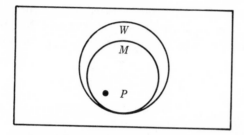

W: Workers
M: Men
P: Patrick

FIGURE 10.2

Care must be taken that the diagrammatic representation of the premises does not represent relationships not required by the premises. The next example illustrates this situation.

EXAMPLE 3.

Consider the argument:

Some students get A's.
Some students are mathematicians.
∴ Some mathematicians get A's.

The relationships expressed by the premises might be represented by either of the diagrams in Figure 10.3. The first diagram indicates that

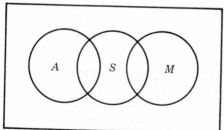 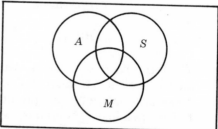

A: Students who get A's   *S*: All students   *M*: Mathematicians

FIGURE 10.3

no mathematicians get A's and the second that some mathematicians get A's; the second diagram expresses a relationship not required by the premises in that there exists a representation of the premises that does not express this relationship. Hence, the conclusion "Some mathematicians get A's" is invalid.

If there is difficulty in determining which of several possible diagrams properly represent the premises, it is helpful to remember that, if there is one representation in which the conclusion is not forced, the argument is invalid.

The form in which the last argument is stated will be the standard form from this point on. The premises are stated first and then the conclusion separated from the premises by a line. The three-dot symbol ∴ is to be read "therefore."

EXAMPLE 4.

All automobiles are manufactured in Germany.
All Chevrolets are automobiles.
_____
∴ All Chevrolets are manufactured in Germany.

Figure 10.4 indicates that all Chevrolets are contained in the set of items manufactured in Germany. The conclusion is valid, but is obviously not true!

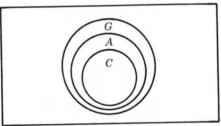

G: German manufactured items
A: Automobiles
C: Chevrolets

FIGURE 10.4

The terms *valid* and *true* as used in logic have different meanings. A true statement is a statement which conforms to reality; validity is concerned only with the question of whether or not a statement follows as a logical consequence from other statements, in this case from the premises—regardless of whether any of the statements are true.

EXERCISES

1. Determine whether or not the conclusion in each of the following arguments is valid or invalid, using Venn diagrams.

   (a)†   All congressmen are honest.
   All senators are congressmen.

       ∴ All senators are honest.

   (b)†   Some magazines are worth reading.
   Collegean is a magazine.

       ∴ *Collegean* is worth reading.

   (c)†   Students who work hard get good jobs.
   Paul is a student who works hard.

       ∴ Paul will get a good job.

   (d)   All mathematicians are ambitious.
   Some students are mathematicians.

       ∴ Some students are ambitious.

   (e)   Some activists are liberals.
   Some congressmen are liberals.

       ∴ Some congressmen are activists.

   (f)†   All understanding parents are good parents.
   Tom's mother is understanding.

       ∴ Tom's mother is a good parent.

   (g)   No even number is divisible by 5.
   30 is an even number.

       ∴ 30 is not divisible by 5.

   (h)   All the "beautiful people" enjoy rock music.
   All humanitarians are "beautiful people."
   Some sociologists are humanitarians.

       ∴ Some sociologists enjoy rock music.

2. Draw a valid conclusion from each of the following sets of premises. Test the validity of your conclusion by using diagrams.

   (a) All department store employees are patient people.
   Some students are department store employees.

   (b) All coeds are beautiful.
   Teresa is not beautiful.

10.2 COMPOUND STATEMENTS

The diagram method of analyzing arguments gives a pictorial representation of how valid conclusions are forced by the premises. Although a satisfactory introduction, the method has some undesirable features. If the number of premises is large, the diagram method becomes unmanageable. Also, care must be taken that unwarranted relationships are not represented. A system which would eliminate these features would be more desirable.

You may have noticed that some of the arguments of the previous section were in exactly the same form. Compare the diagrams for Examples 1 and 4 with that of Exercise 1(a); compare the diagrams for Example 2, Exercise 1(c), and Exercise 1(f). These comparisons indicate that it is the form of the argument that determines validity, not the content of any of the statements.

Symbolic logic, as we shall see shortly, does in fact consider the form of an argument, rather than the content, and eliminates the undesirable features of diagram analysis that were pointed out above.

Previously we made no formal definition of the word *statement*. We must do so now to develop symbolic logic in a precise manner. A *statement* shall be a declarative sentence that is either true (T) or false (F), but not both; the truth or falsity of a statement is its *true value*.

Of the following sentences, the first two are not statements; the last two are statements:

> What time is it?
> Stop when I tell you.
> Tokyo is the largest city in the world.
> Every house has a doorway.

Sentences that are not statements have no role in symbolic logic.

Statements will be denoted by lower case letters, usually by p, q, r, or s. Designations such as

> p: Tokyo is the largest city in the world.

indicate both the content of a statement and the letter which denotes the statement.

In ordinary language usage, several statements are often combined to form compound statements; hence, the study of logic requires the consideration of compound statements. The three types of compound statements to be considered are *conjunction*, *disjunction*, and *implication*. These types of compound statements and their notation are illustrated in Table 10.1, using the simple statements

> p: Mathematics is a tool of science.
> q: Mathematics is a college requirement.

TABLE 10.1

	Statement	*Connective*	*Symbolically*
Conjunction	Mathematics is a tool of science and mathematics is a college requirement.	and	$p \wedge q$
Disjunction	Mathematics is a tool of science or mathematics is a college requirement.	or	$p \vee q$
Implication	If mathematics is a tool of science, then mathematics is a college requirement.	if . . . , then . . .	$p \rightarrow q$

These connectives in turn can be used in combination with each other. For example, if p denotes "The feminist movement has merit," q denotes "Ann is involved in politics," and r denotes "Newspapers are necessary for a democracy," then

$(p \wedge q) \vee r$	denotes	"Either the feminist movement has merit and Ann is involved in politics, or newspapers are necessary for democracy."
$(p \vee r) \rightarrow q$	denotes	"If the feminist movement has merit or newspapers are necessary for a democracy, then Ann is involved in politics."
$(p \wedge q) \wedge r$	denotes	"The feminist movement has merit and Ann is involved in politics, and newspapers are necessary for a democracy."
$p \rightarrow (q \vee r)$	denotes	"If the feminist movement has merit, then Ann is involved in politics or newspapers are necessary for a democracy."

The *negation* of a statement p is denoted by the symbol $\sim p$. The negations and the symbolic representations of the three statements p, q, and r above are

$\sim p$: The feminist movement does not have merit.
$\sim q$: Ann is not involved in politics.
$\sim r$: Newspapers are not necessary for a democracy.

Negation is also used in combination with the compound connectives. Continuing with the above p, q, and r,

$(\sim p) \vee q$	denotes	"The feminist movement does not have merit or Ann is involved in politics."
$\sim(p \vee q)$	denotes	"It is not true that either the feminist movement has merit or Ann is involved in politics."

$p \to (\sim r)$ denotes   "If the feminist movement has merit, then newspapers are not necessary for a democracy."

It was indicated above that an implication is a statement of the form "if p, then q." In ordinary language usage, implications are often stated in different forms. The following is a list of some such forms, each of which is expressed symbolically by $p \to q$:

if p, then q
p is a sufficient condition for q
q is a necessary condition for p
p only if q
q if p.

For example, if p denotes "I study history" and q denotes "I enjoy history," the statement "I study history only if I enjoy history" is denoted by $p \to q$. "That I enjoy history is a necessary condition that I study history" is similarly represented.

There is a fourth type of compound statement that occurs in mathematics. This is the *biconditional*, recognized by the connective "if and only if" (or "is a necessary and sufficient condition"). The symbol for this connective is the double arrow, $\longleftrightarrow$.

If, as above, p denotes "The feminist movement has merit," and q denotes "Ann is involved in politics," then $p \longleftrightarrow q$ denotes "The feminist movement has merit if and only if Ann is involved in politics" (alternately, "That the feminist movement has merit is a necessary and sufficient condition that Ann is involved in politics").

Because the biconditional can be replaced by a combination of implications and a conjunction, it will not be used further here.

EXERCISES

1.† Give the symbolic form of the following sentences, using

   p: Exercising is essential for good health.
   q: Eating too much causes strain on the heart.
   r: Good health retards old age.

(a)   Eating too much does not cause strain on the heart.
(b)   Exercising is essential for good health and good health retards old age.
(c)   If exercising is not essential for good health, then good health does not retard old age.
(d)   Eating too much causes strain on the heart but good health does not retard old age.
(e)   Exercising is essential for good health or eating too much causes strain on the heart, or good health retards old age.

2. Give the symbolic form of the following sentences, using

   P: A liberal education broadens the intellectual horizons of a student.
   q: Mathematics is part of our intellectual climate.
   r: Mathematics is part of a liberal education.

   (a) Mathematics is not part of a liberal education.
   (b) If a liberal education broadens the intellectual horizons of a student and mathematics is part of our intellectual climate, then mathematics is part of a liberal education.
   (c) That a liberal education broadens the intellectual horizons of a student is a sufficient condition that mathematics is a part of a liberal education.
   (d) If mathematics is not part of a liberal education, then a liberal education does not broaden the intellectual horizons of the student and mathematics is not part of our intellectual climate.
   (e) That mathematics is part of our intellectual climate is a necessary condition that mathematics is part of a liberal education.

3.† If *p*, *q*, and *r* denote the statements indicated in Exercise 1, write in words the sentences expressed symbolically by:

   (a) $(p \wedge q) \rightarrow r$.   (b) $(\sim p) \vee (\sim r)$.
   (c) $(p \rightarrow r) \wedge (r \rightarrow q)$.   (d) $(\sim q) \vee r$.

4. If *p*, *q*, and *r* denote the statements indicated in Exercise 2, write in words the sentences expressed symbolically by:

   (a) $[(\sim p) \wedge (\sim q)] \wedge (\sim r)$.   (b) $\sim p \wedge q$.
   (c) $(\sim r) \rightarrow (p \wedge q)$.   (d) $(p \vee r) \wedge (\sim q)$.

10.3 TRUTH TABLES AND EQUIVALENT STATEMENTS

As we shall see later, valid conclusions depend on the truth values of the premises and of the conclusion. Because the premises and the conclusion may be simple or compound statements, of particular interest are the truth values of compound statements as determined by the truth values of their component parts. The usual method of determining truth values is to use Tables 10.2, 10.3, 10.4, and 10.5, called *truth tables*.

TABLE 10.2

p	*q*	$p \wedge q$
T	T	T
T	F	F
F	T	F
F	F	F

TABLE 10.3

p	*q*	$p \vee q$
T	T	T
T	F	T
F	T	T
F	F	F

TABLE 10.4

p	*q*	$p \rightarrow q$
T	T	T
T	F	F
F	T	T
F	F	T

TABLE 10.5

p	$\sim p$
T	F
F	T

Table 10.2 is read as follows: The conjunction of p and q is true when both p and q are true; the conjunction is false when p is true and q false, when p is false and q is true, and when both p and q are false. The other tables are read similarly. Note that all possible combinations of truth values of p and q are considered. Note also that the conjunction of two statements is true only when both are true, that the disjunction is false only when both statements are false, and that the implication $p \rightarrow q$ is false only when p (the *antecedent*) is true and q (the *consequent*) is false.

From the above form tables, the truth values for more complex statements can be determined. Table 10.6 is the truth table for the statement $(p \vee q) \rightarrow (\sim q)$. The entries in Column 3 are based on Table 10.3; those in Column 4 are based on Table 10.5 by considering the entries in Column 2; and those in Column 5 are based on Table 10.4 by comparing the entries

TABLE 10.6

1	*2*	*3*	*4*	*5*
p	q	$p \vee q$	$\sim q$	$(p \vee q) \rightarrow (\sim q)$
T	T	T	F	F
T	F	T	T	T
F	T	T	F	F
F	F	F	T	T

in Columns 3 and 4, treating $p \vee q$ as the antecedent and $\sim q$ as the consequent. (It is important to realize that the statements p and q in the tables can be compound statements or even negations.) Column 5 gives the truth values of $(p \vee q) \rightarrow (\sim q)$ for all possible combinations of truth values of p and q!

As another example, consider the truth table for the statement $\sim(p \vee q) \wedge q$ (Table 10.7). The entries in Column 4 are based on Table 10.5 by considering the entries in Column 3; the entries in Column 5 are based on Table 10.2 by considering the entries in Columns 2 and 4. Again

TABLE 10.7

1	*2*	*3*	*4*	*5*
p	q	$p \vee q$	$\sim(p \vee q)$	$\sim(p \vee q) \wedge q$
T	T	T	F	F
T	F	T	F	F
F	T	T	F	F
F	F	F	T	F

Column 5 gives the truth values of the statement $\sim(p \vee q) \wedge q$ for all possible combinations of truth values of p and q.

The statement $(p \wedge q) \rightarrow (\sim r)$ is made up of the three component statements p, q, and r. To consider all combinations of the truth values of p, q, and r requires a truth table with eight rows (Table 10.8). The entries

TABLE 10.8

1	2	3	4	5	6
p	q	r	$p \wedge q$	$\sim r$	$(p \wedge q) \rightarrow (\sim r)$
T	T	T	T	F	F
T	T	F	T	T	T
T	F	T	F	F	T
T	F	F	F	T	T
F	T	T	F	F	T
F	T	F	F	T	T
F	F	T	F	F	T
F	F	F	F	T	T

in Column 4 are based on Table 10.2 by considering Columns 1 and 2; those in Column 5 on Table 10.5 by considering Column 3; and those in Column 6 on Table 10.4 based on Columns 4 and 5. In this case, Column 6 gives the truth values of $(p \wedge q) \rightarrow (\sim r)$ for all possible combinations of truth values of p, q, and r.

EXERCISES

Construct truth tables for the following statements.

1.† $p \wedge (\sim q)$
3.† $(\sim p) \wedge (\sim q)$
5. $\sim(\sim q)$
7. $(p \vee q) \wedge [\sim(p \rightarrow q)]$
9.† $r \wedge (p \rightarrow q)$

2.† $\sim(p \vee q)$
4. $[p \vee (\sim q)] \rightarrow q$
6.† $\sim(p \rightarrow q)$
8. $[p \wedge (q \vee \sim p)] \rightarrow (\sim q)$
10. $(\sim r) \rightarrow (p \rightarrow q)$

Table 10.9 combines the truth tables of $\sim(p \wedge q)$ and $(\sim p) \vee (\sim q)$. A comparison of the fourth and seventh columns of the truth table shows that the two statements, $\sim(p \wedge q)$ and $(\sim p) \vee (\sim q)$, always have the same truth values. Such statements are called equivalent statements. In general, two statements are *equivalent statements* if they always have the same truth values. In establishing the validity of arguments later in this chapter, it will often be convenient to replace a statement by an equivalent statement.

TABLE 10.9

p	q	$p \wedge q$	$\sim(p \wedge q)$	$\sim p$	$\sim q$	$(\sim p) \vee (\sim q)$
T	T	T	F	F	F	F
T	F	F	T	F	T	T
F	T	F	T	T	F	T
F	F	F	T	T	T	T

Some of the pairs of equivalent statements most frequently used are given in Table 10.10. The first three pairs give equivalent forms for the

TABLE 10.10   Equivalent Statements

Statement	Equivalent
1. $\sim(p \wedge q)$	$(\sim p) \vee (\sim q)$
2. $\sim(p \vee q)$	$(\sim p) \wedge (\sim q)$
3. $\sim(p \rightarrow q)$	$p \wedge (\sim q)$
4. $p \rightarrow q$	$(\sim p) \vee q$
5. $p \rightarrow q$	$(\sim q) \rightarrow (\sim p)$
6. $\sim(\sim q)$	q
7. $p \wedge q$	$q \wedge p$
8. $p \vee q$	$q \vee p$

negation of the basic compound statements. Pairs 4 and 5 give alternate forms of the implication. Pair 6 is the double negation. Pairs 7 and 8 are commutative laws for conjunction and disjunction.

That the two statements in Pair 1 are equivalent is verified by the truth table in Table 10.9. That the two statements in Pair 2 are equivalent can be verified by comparing Exercises 2 and 3 of the previous set of exercises in this section; that the two statements in Pair 3 are equivalent can be verified by comparing Exercises 1 and 6 of the previous set of exercises. The verification that some of the remaining pairs of statements are equivalent is left to the exercises at the end of this section.

It is important to realize that the statements p and q in the given pairs of equivalent statements can be either simple, compound, or negations. For example, the two statements

$$\sim[p \rightarrow (s \vee t)] \quad \text{and} \quad p \wedge [\sim(s \vee t)]$$

are equivalent statements in the form of the third pair of statements in Table 10.10, with q being in this case $(s \vee t)$.

Similarly,

$$\sim[s \vee (p \rightarrow q)] \quad \text{and} \quad (\sim s) \wedge [\sim(p \rightarrow q)]$$

are in the second form of the listing with p being s and q being $p \rightarrow q$.

EXERCISES

1.[†] Verify that the statements in Pair 4, Table 10.10, are equivalent.

2. Verify that the statements in Pair 5, Table 10.10, are equivalent.

3.[†] Verify that the statements in Pair 6, Table 10.10, are equivalent.

4. The truth table for the biconditional (page 322) is shown in Table 10.11.

TABLE 10.11

p	q	$p \longleftrightarrow q$
T	T	T
T	F	F
F	T	F
F	F	T

Verify that $p \longleftrightarrow q$ is equivalent to $(p \to q) \land (q \to p)$.

5. Each of the following pairs of statements are equivalent because they are in the form of one of the pairs of equivalent statements in Table 10.10. Determine into which of these forms each pair falls.

(a)† $s \to t$ $(\sim t) \to (\sim s)$
(b) $\sim[p \lor (s \land t)]$ $(\sim p) \land [\sim(s \land t)]$
(c)† $\sim[\sim(s \to t)]$ $s \to t$
(d) $\sim[p \land (p \to q)]$ $(\sim p) \lor [\sim(p \to q)]$
(e)† $\sim[(p \lor q) \to q]$ $(p \lor q) \land (\sim q)$
(f) $(p \land q) \to (p \lor q)$ $[\sim(p \land q)] \lor (p \lor q)$

10.4   APPLICATION OF LOGIC TO SWITCHING CIRCUITS

This section, although a temporary diversion from our central purpose of studying valid arguments, considers a remarkable area of application of symbolic logic. The area of application is in the design of switching circuits. These circuits are as commonplace as the electrical circuits that light your house and as sophisticated as the arithmetic and logical circuits in a computer. No knowledge of electricity is required in our study.

A switch is a mechanical device that allows electricity to flow between two points, A and B, when closed (Fig. 10.5) or stops the flow when

A B

FIGURE 10.5

the switch is open (Fig. 10.6). We are not concerned with the mechanical device itself, only the manner in which switches are combined to form more complex networks.

FIGURE 10.6

Hereafter, instead of illustrating switches as in Figures 10.5 and 10.6, we shall designate a switch by a letter as in Figure 10.7. We will refer to p

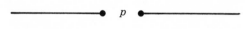

FIGURE 10.7

as being closed to allow the flow of electricity or open to stop the flow. The endpoints, A and B, of the circuit will also be omitted in the Figures.

The two basic ways in which switches are connected are in *series* (Fig. 10.8) and in *parallel* (Fig. 10.9). For electricity to flow through the

FIGURE 10.8

circuit when connected in series, both p and q must be closed. If either one is open, electricity will not flow from the left endpoint to the right.

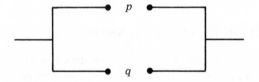

FIGURE 10.9

In the parallel connection, electricity will flow through the circuit when either p or q (or both) is closed. When both are open there is no flow.

Tables 10.12 and 10.13 give the states of the circuits for the various states of p and q. T denotes the switch is closed or electricity is flowing;

TABLE 10.12

p	q	Series
T	T	T
T	F	F
F	T	F
F	F	F

TABLE 10.13

p	q	Parallel
T	T	T
T	F	T
F	T	T
F	F	F

F denotes the switch is open or electricity is not flowing. Do these tables look familiar? The result is that the mathematical model for a series connection is a conjunction, $p \wedge q$; the mathematical model for a parallel connection is a disjunction, $p \vee q$!

Before examining how these models are used, we shall consider how the two basic connections, series and parallel, are used to form more complex circuits. An example is shown in Figure 10.10. In this circuit, p and q are connected in parallel. This combination is then connected in series

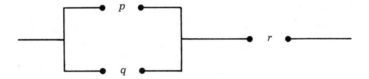

FIGURE 10.10

with switch r. Its mathematical model, hereafter called its *logical representation*, is $(p \vee q) \wedge r$. For electricity to flow, r must be open and in addition at least one of p and q.

In some situations, a single switch performs several jobs and therefore has several connections—for example, when the light switch in a bathroom not only turns on the light but also the exhaust fan. Such switches will appear several times in our diagrams, but always with the same letter. Similarly for two switches that are always open or closed together. Otherwise, they receive no special attention.

EXAMPLE 1.

Determine the logical representation of the circuit in Figure 10.11.

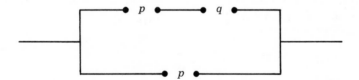

FIGURE 10.11

In this circuit, p and q are first connected in series. The combination is then connected in parallel with p. Its logical representation is $(p \wedge q) \vee p$.

Finally, two switches can be so interconnected that when one is open the other is closed, and vice versa—for example, the switches that control the blinking lights at a railroad crossing; when one light is on, the other is off. In our diagrams, two such switches will be designated by the same letter, one with a negation sign, for example, p and $\sim p$.

EXAMPLE 2.

Determine the logical representation of the circuit in Figure 10.12.

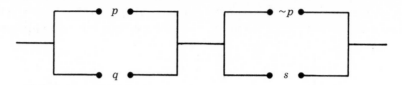

FIGURE 10.12

The logical representation is $(p \vee q) \wedge (\sim p \vee s)$. For electricity to flow through this circuit, at least one of p and q, and at least one of $\sim p$ and s, must be closed. However, because of the relationship between p and $\sim p$, the only possible combinations which allow the flow are p and s, or q and $\sim p$.

EXAMPLE 3.

Draw the circuit with the logical representation $(p \vee q) \vee (s \wedge \sim p)$.

The groups in parentheses are p and q in parallel (Fig. 10.13) and s and $\sim p$ in series (Fig. 10.14).

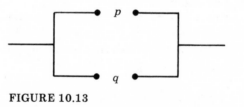

FIGURE 10.13

FIGURE 10.14

These two groups are then connected in parallel (Fig. 10.15).

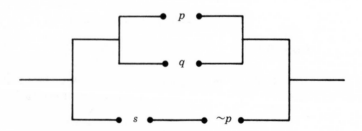

FIGURE 10.15

EXERCISES

1. Under what conditions will electricity flow through the circuits in Figures 10.16 through 10.21?

    (a)†

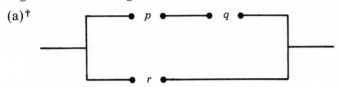

    FIGURE 10.16

    (b)

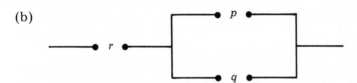

    FIGURE 10.17

    (c)†

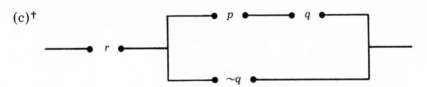

    FIGURE 10.18

    (d)

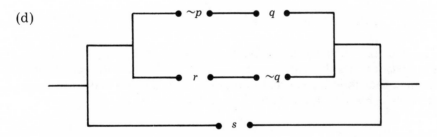

    FIGURE 10.19

    (e)† ———● *p* ●——————● ~*p* ●———

    FIGURE 10.20

(f)

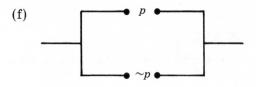

FIGURE 10.21

2.† Give the logical representation of each of the circuits in Exercise 1.

3. Draw the circuit with the following representations:

(a)† $p \lor (q \land s)$ (b) $p \land (q \lor s)$

(c)† $(p \lor q) \land (\sim p \lor q)$ (d) $p \lor [(q \land \sim p) \lor s]$

(e)† $p \land [(q \lor r) \land s]$

(f) $[(p \land q) \lor (\sim p \land \sim q)] \land [s \land (p \lor [\sim p \land s])]$

Which of the circuits in Figure 10.22 would cost less to manufacture?

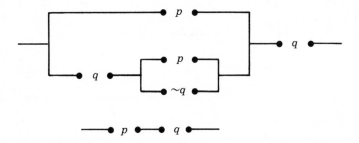

FIGURE 10.22

The two are equivalent in the sense that, when electricity flows through the first, it flows through the second, and vice versa. This corresponds to the fact that their logical representation, whose combined truth tables are given in Table 10.14, are equivalent statements.

TABLE 10.14

p	q	$(p \lor [q \land (p \lor \sim q)]) \land q$	$p \land q$
T	T	T	T
T	F	F	F
F	T	F	F
F	F	F	F

Because the two circuits are equivalent, either can be used in place of the other. Manufacturers would normally prefer the simpler circuit at the bottom because of the reduced cost. The savings may only be a few cents, but if thousands of circuits are involved, the savings could be significant. Our objective is to consider a process of simplifying a given circuit by reducing its logical representation to a simpler but equivalent form.

The list of additional pairs of equivalent statements in Table 10.15 will be useful for this prupose. The numbering continues from that of Table 10.10.

TABLE 10.15 Equivalent Statements

	Statement	*Equivalency*	
7.	$p \wedge q$	$q \wedge p$	$\left.\right\}$ Commutative Laws
8.	$p \vee q$	$q \vee p$	
9.	$p \wedge (q \wedge r)$	$(p \wedge q) \wedge r$	$\left.\right\}$ Associative Laws
10.	$p \vee (q \vee r)$	$(p \vee q) \vee r$	
11.	$p \wedge (q \vee r)$	$(p \wedge q) \vee (p \wedge r)$	$\left.\right\}$ Distributive Laws
12.	$p \vee (q \wedge r)$	$(p \vee q) \wedge (p \vee r)$	
13.	$p \wedge (q \vee {\sim}q)$	p	
14.	$p \vee (q \wedge {\sim}q)$	p	
15.	$p \wedge (p \vee q)$	p	
16.	$p \vee (p \wedge q)$	p	

Again, when using the pairs of equivalent statements in Table 10.15, it is essential to keep in mind that p, q, and r can represent compound statements or negations.

EXAMPLE 4.

Simplify the circuit $(p \wedge q) \vee (p \wedge {\sim}q)$.

The given representation is in the form of the statement on the right of Pair 11 in the listing. Its equivalent form from the left is

$$p \wedge (q \vee {\sim}q),$$

which from Pair 13 in the listing is equivalent to p. The given circuit simplifies to one consisting of a single switch p.

EXAMPLE 5.

Simplify the circuit in Figure 10.23.

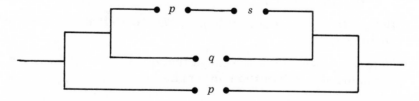

FIGURE 10.23

The logical representation is

$$p \vee [(p \wedge s) \vee q],$$

which by Pair 10 of Table 10.15 is equivalent to

$$[p \vee (p \wedge s)] \vee q$$

and by Pair 16 is equivalent to $p \vee q$.

EXAMPLE 6.

Simplify the circuit in Figure 10.24.

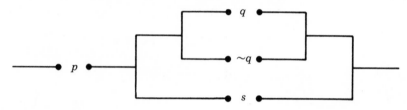

FIGURE 10.24

The logical representation is

$$p \wedge [(q \vee \sim q) \vee s],$$

which by Pair 11 of Table 10.15 is equivalent to

$$[p \wedge (q \vee \sim q)] \vee (p \wedge s)$$

and by Pair 13 is equivalent to

$$p \vee (p \wedge s)$$

and finally by Pair 16 to p.

EXAMPLE 7.

Simplify the circuit $(p \vee [q \wedge (p \vee \sim q)]) \wedge q$.

By Pair 11 of Table 10.15, $q \wedge (p \vee \sim q)$ is equivalent to $(q \wedge p)$ $\vee (q \wedge \sim q)$, which by Pair 13 is equivalent to $(q \wedge p)$, or by Pair 7 to $(p \wedge q)$. Substituting $(p \wedge q)$ into the given representation gives

$$[p \vee (p \wedge q)] \wedge q.$$

But $p \vee (p \wedge q)$ is equivalent to p by Pair 16, so that the simplified circuit is just $p \wedge q$.

Note that this circuit is the one at the top in Figure 10.22 in the beginning of this discussion on simplifying circuits.

EXERCISES

1.† Construct a truth table to verify that the statements in Pair 9, Table 10.15, are equivalent.

2. Construct a truth table to verify that the statements in Pair 11, Table 10.15, are equivalent.

3. Determine the logical representation and then simplify the circuits in Figures 10.25 through 10.29.

(a)†

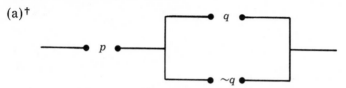

FIGURE 10.25

(b)

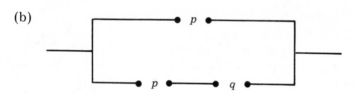

FIGURE 10.26

(c)†

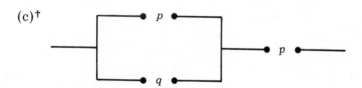

FIGURE 10.27

(d)

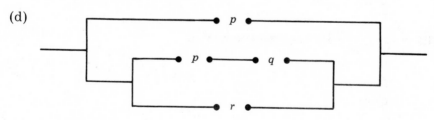

FIGURE 10.28

(e)†

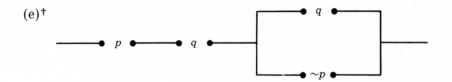

FIGURE 10.29

4. Simplify the following circuits:

(a)† $(p \wedge q) \vee (\sim p \wedge q)$.

(b) $p \vee [(q \wedge s) \vee (q \wedge \sim s)]$.

(c)† $p \wedge [q \wedge (\sim q \vee r)]$.

(d) $[(p \wedge q) \vee (\sim p \wedge q)] \wedge (p \vee q)$.

10.5 VALID ARGUMENTS

What are we doing? Our pursuit of logic has been for the purpose of determining whether an argument is valid, that is, whether a statement is a logical consequence of the premises. This determination, in turn, is for the purpose of analyzing mathematical proofs to see why a proof does what it pretends to do, namely to establish the validity of a conclusion (a theorem).

To do this, we must determine exactly what are valid arguments and valid conclusions. Before doing so, consider the following argument:

> If Dan is a student, he studies hard.
> Dan is a student.
> ∴ Dan studies hard.

We can use the representation

p: Dan is a student.
q: Dan studies hard.

Then, the above argument can be written

> $p \rightarrow q$
> p
> ∴ q

or, finally,

$$[(p \rightarrow q) \wedge p] \rightarrow q.$$

This last form states that the conjunction of the premises implies the conclusion. If this argument is to be valid, it seems reasonable to require that this last implication be always true. Consider its truth table in Table 10.16.

TABLE 10.16

p	q	$p \to q$	$(p \to q) \land p$	$[(p \to q) \land p] \to q$
T	T	T	T	T
T	F	F	F	T
F	T	T	F	T
F	F	T	F	T

The last column indicates that it is the case that the implication is always true, that the argument would be valid and therefore that "Dan studies hard" would be a valid conclusion.

In general, *an argument is valid if the statement that the conjunction of the premises implies the conclusion is always true*. In this case, the conclusion is said to be a *valid conclusion*. In the truth table of Table 10.16, note that the conclusion q can at times be false, despite the fact that the argument is valid!

EXAMPLE 1.

For an example of an invalid argument, consider

> Either Tim is a barber or barbers talk too much.
> Barbers do not talk too much.
> _____
> ∴ Tim is not a barber.

We use the representation

> p: Tim is a barber.
> q: Barbers talk too much.

The argument in symbolic form becomes

> $p \lor q$
> $\sim q$
> _____
> ∴ $\sim p$

and in implication form

> $[(p \lor q) \land (\sim q)] \to (\sim p)$.

The question of validity depends on whether or not this last implication is always true (Table 10.17).

TABLE 10.17

p	q	$(p \lor q)$	$\sim q$	$(p \lor q) \land (\sim q)$	$\sim p$	$[(p \lor q) \land (\sim q)] \to (\sim p)$
T	T	T	F	F	F	T
T	F	T	T	T	F	F
F	T	T	F	F	T	T
F	F	F	T	F	T	T

Because this implication is not always true, the argument is invalid.

EXERCISES

Determine whether each of the following arguments is valid or invalid. In Exercises 1 through 5, first convert the argument into symbolic form.

1.†   If Paul studies programming, he could use the computer.
    Paul cannot use the computer.

  ∴ Paul does not study programming.

2.   If I have a job, I have a car.
    If I have a car, I must buy expensive insurance.

  ∴ If I have a job, I must buy expensive insurance.

3.   Winning isn't everything or losing is nothing.
    Losing does have some merit (losing is something).

  ∴ Winning is everything.

4.†   Elected representatives serve their constituents or they lose their jobs.
    But elected representatives do not lose their jobs.

  ∴ Elected representatives do not serve their constituents.

5.†   Either Joan does not care for people or she studies sociology.
    Joan does care for people.

  ∴ Joan studies sociology.

6.† $p \to q$ 7. $p \lor q$ 8.† $p \to q$
     $\sim q$ $\sim p$ $q \to r$

  ∴ $\sim p$ ∴ q ∴ $p \to r$

9. $p \land q$ 10.† $p \to q$
     p p

  ∴ q ∴ q

10.6 RULES OF INFERENCE

One conclusion to be drawn from Section 10.5 is that the validity of an argument is dependent on the form of the argument, rather than on the content of any of the statements.

Compare the symbolic form of the valid argument given in the first argument of Section 10.5 with that of Exercise 10 of that section; compare the symbolic form of the valid argument in Exercise 1 with that of Exercise 6; the symbolic form of Exercise 2 with that of Exercise 8; that of Exercise 5 with Exercise 7. Once an argument form has been verified to be valid, there is no purpose in repeating the verification each time it is encountered. That is, we can compile a list of verified argument forms and compare any given argument against this listing to verify its validity. Such argument forms are called *rules of inference*.

Some of the most commonly encountered argument forms are given in Table 10.18.

TABLE 10.18 Rules of Inference

1. $p \to q$
       p
     $\therefore\ q$

2. $p \to q$
       $q \to r$
     $\therefore\ p \to r$

3. $p \to q$
       $\sim q$
     $\therefore\ \sim p$

4. $p \lor q$ or $p \lor q$
       $\sim p$ $\sim q$
     $\therefore\ q$ $\therefore\ p$

5. $p \land q$ or $p \land q$
     $\therefore\ p$ $\therefore\ q$

6. p or q
     $\therefore\ p \lor q$ $\therefore\ p \lor q$

7. p or q
       q p
     $\therefore\ p \land q$ $\therefore\ p \land q$

The validity of Rule 5 of Table 10.18 is established by the truth table in Table 10.19. The last two columns indicate that the two forms of Rule 5 are valid.

TABLE 10.19

p	q	$p \wedge q$	$(p \wedge q) \to p$	$(p \wedge q) \to q$
T	T	T	T	T
T	F	F	T	T
F	T	F	T	T
F	F	F	T	T

The verification that Rules 1, 2, and 3 are also valid was done in Exercises 10, 8, and 6, respectively, of Section 10.5. The verification that Rules 4, 6, and 7 are valid is left to the exercises.

When using the rules of inference, it is again necessary to keep in mind that p and q can be compound statements or negations. For example, the argument

$$(s \wedge t) \to q$$
$$\underline{\sim q}$$
$$\therefore \; \sim (s \wedge t)$$

is valid because it is in the form of Rule 3, with the statement p of Rule 3 being $(s \wedge t)$ in this instance.

Consider the argument

$$\underline{\sim (t \to u)}$$
$$\therefore \; [\sim (t \to u)] \vee s,$$

which is valid because it is in the first form of Rule 6, with the statement p of Rule 6 being $\sim (t \to u)$ and the statement q of Rule 6 being s.

Finally,

$$(p \wedge q) \to q$$
$$\underline{q \to r}$$
$$\therefore \; (p \wedge q) \to r$$

is valid because it is in the form of Rule 2, with the statement p of Rule 2 being $(p \wedge q)$ in this argument.

EXERCISES

1.† Verify that Rule 4, Table 10.18, is a rule of inference.

2. Verify that Rule 6, Table 10.18, is a rule of inference.

3.† Verify that Rule 7, Table 10.18, is a rule of inference.

4. Each of the following arguments is valid because it falls into the form of one of the rules of inference in Table 10.18. Determine into which form each falls.

(a)†
$$s \lor (\sim t)$$
$$\sim(\sim t)$$
$$\therefore s$$

(b)
$$p \to (r \lor s)$$
$$p$$
$$\therefore r \lor s$$

(c)†
$$r \to s$$
$$q$$
$$\therefore (r \to s) \land q$$

(d)
$$\sim r$$
$$\therefore (\sim p) \lor (\sim r)$$

(e)†
$$p \to (p \lor q)$$
$$\sim(p \lor q)$$
$$\therefore \sim p$$

(f)
$$(\sim p) \land (p \land q)$$
$$\therefore p \land q$$

(g)†
$$p \to (\sim q)$$
$$(\sim q) \to (\sim r)$$
$$\therefore p \to (\sim r)$$

(h)
$$p \lor (s \to q)$$
$$\sim p$$
$$\therefore s \to q$$

10.7 ANALYSIS OF VALID ARGUMENTS BY RULES OF INFERENCE

That an argument is valid can now be verified by use of rules of inference, without the use of truth tables, although the determination that an argument is invalid must still be done by truth tables. In the use of rules of inference, any statement may be replaced by an equivalent statement when it is convenient to do so. Because equivalent statements always have the same truth values, this will not affect the validity of an argument.

Usually several rules of inference are used in any valid argument. Therefore, the format of the analyses of arguments that follow will give a listing of valid statements which lead to the desired conclusion. With each statement will be the reason(s) why that statement is valid. The reasons will be either a premise (postulate, axiom) (abbreviated P1, P2, P3, . . .), a rule of inference from Table 10.18 (abbreviated R1, R2, R3, . . .), a statement already established as valid (abbreviated S1, S2, S3, . . .), an indication that a statement is equivalent to a premise or a statement that is already established as valid (abbreviated ES), or a combination of these four types of reasons.

The following examples illustrate this method of verifying that arguments are valid.

EXAMPLE 1.

> If automobiles pollute, they are modified.
> If automobiles are modified, the cost of living increases.
> The cost of living does not increase.
> Therefore, automobiles do not pollute.

If we symbolize as follows:

> s: Automobiles pollute
> t: Automobiles are modified
> u: The cost of living increases,

the argument can be written

> Premise 1: $s \rightarrow t$
> Premise 2: $t \rightarrow u$
> Premise 3: $\sim u$
> _____
> $\therefore$ $\sim s$.

Proof:

Statement Number	Valid Statement	Reasons
1	$\sim t$	P2, P3, R3
2	$\sim s$	S1, P1, R3

Because $\sim s$ is obtained as a valid conclusion, the argument is valid.

Note the reasons for Statement 1; by Rule 3,

> $t \rightarrow u$
> $\sim u$
> _____
> $\therefore$ $\sim t$

is a valid argument (with t in place of p and u in place of q); the validity of $\sim u$ follows from Premise 3 and that of $t \rightarrow u$ from Premise 2. Because all three (P2, P3, and R3) are essential ingredients in the validity of Statement 2, each must be indicated as a reason. The use of Rule 3 requires that $\sim u$ and $t \rightarrow u$ be first established as valid; P2 and P3 indicate where this validity is established.

In a similar manner, Statement 3 is valid by Rule 3 from Premise 1 and Statement 1. For practice, you should construct the argument.

EXAMPLE 2.

> Triangle ABC is an isosceles triangle. If angle A is acute, then angle B is acute. Angle B is not acute. Therefore, angle A is not acute or triangle ABC is not isosceles.

If we symbolize as follows:

   s: Triangle *ABC* is an isosceles triangle
   t: Angle *A* is acute
   u: Angle *B* is acute,

the argument can be written

   P1: *s*
   P2: *t* → *u*
   P3: ~*u*
   ∴ (~*t*) ∨ (~*s*).

Proof:

Statement Number	Valid Statement	Reasons
1	(~*u*) → (~*t*)	P2, ES
2	~*t*	S1, P3, R1
3	(~*t*) ∨ (~*s*)	S2, R6

Because (~*t*) ∨ (~*s*) follows as a valid conclusion, the argument is valid.

Statement 1 in this example is valid because it is an equivalent form of Premise 2 (Pair 5 of equivalent statements, Table 10.10).
   Statement 2 is valid because the argument

   (~*u*) → (~*t*)   (S1)
   ~*u*   (P3)
   ∴ ~*t*

is valid by Rule 1.

   Statement 3 is valid because the argument

   ~*t*   (S2)
   ∴ (~*t*) ∨ (~*s*)

is valid by Rule 6. The reasons given for the validity of each of the three statements in the proof simply reflect all of the ingredients that go into the verification of the particular statements.
   Note that each statement given in the proof is valid because it fits into one of two classifications:

1. The statement is equivalent to a premise or to a previous statement.
2. The statement is the conclusion of a valid argument that can be constructed, according to one of the rules of inference, from the premises and/or prior statements.

All of the statements given in the proofs of arguments in this section will be one of these two types. The next example involves the use of Rule 3 when the q of Rule 3 is compound; also involved are several negations.

EXAMPLE 3.

Inflation rises and prices increase when demand exceeds supply. Prices are not rising. Therefore, demand does not exceed supply.

Symbolically:

r: Demand exceeds supply
s: Inflation rises
t: Prices increase

P1: $r \to (s \wedge t)$
P2: $\sim t$
_____
$\therefore \sim r.$

Proof:

Statement Number	Valid Statement	Reasons
1	$(\sim s) \vee (\sim t)$	P2, R6
2	$\sim (s \wedge t)$	S1, ES
3	$\sim r$	S2, P1, R3

Argument valid.

You should construct the arguments by which Statements 1 and 3 are valid and should consider which of the forms in Table 10.10 validates Statement 2 as equivalent to Statement 1.

EXERCISES

Give the reasons why the given statements in the verification of the following arguments are valid.

	Statement Number	Valid Statement
1.† P1: $(\sim p) \to q$	1	$\sim(\sim p)$
P2: $\sim q$	2	p
$\therefore p$		
2. P1: $p \to (\sim q)$	1	$(\sim r) \to \sim(\sim q)$
P2: $(\sim q) \to r$	2	$\sim(\sim q) \to (\sim p)$
$\therefore (\sim r) \to (\sim p)$	3	$(\sim r) \to (\sim p)$

	Statement Number	*Valid Statement*
	or	
	1	$p \to r$
	2	$(\sim r) \to (\sim p)$

3.† P1: $(p \lor q) \to s$ 1 $p \lor q$
   P2: $\underline{p\qquad\qquad}$ 2 s
   ∴ s

4. P1: $(p \land q) \to s$ 1 $\sim(p \land q)$
   P2: $\sim s$ 2 $(\sim p) \lor (\sim q)$
   P3: $\underline{p\qquad\qquad}$ 3 $\sim(\sim p)$
   ∴ $\sim q$ 4 $\sim q$

5.† P1: $p \land q$ 1 p
   P2: $p \to s$ 2 s
   P3: $s \to t$ 3 t
   ∴ t or
     1   p
     2     $p \to t$
     3   t

6. P1: $p \lor q$ 1 $\sim p$
   P2: $r \to (\sim p)$ 2 q
   P3: $\underline{r\qquad\qquad}$
   ∴ q

7.† P1: $(\sim r) \lor q$ 1 $\sim(\sim r)$
   P2: $q \to p$ 2 q
   P3: $\underline{r\qquad\qquad}$ 3 p
   ∴ p

8. P1: $(p \lor q) \to (q \lor s)$ 1 $p \lor q$
   P2: $(\sim r) \land (\sim q)$ 2 $q \lor s$
   P3: $\underline{p\qquad\qquad\qquad\quad}$ 3 $\sim q$
   ∴ s 4 s

9.† P1: $p \to (\sim q)$ 1 p
   P2: $r \to (\sim p)$ 2 $\sim q$
   P3: $\underline{p \land s\qquad\quad}$ 3 $\sim(\sim p)$
   ∴ $\sim(q \lor r)$ 4 $\sim r$
     5     $(\sim q) \land (\sim r)$
     6     $\sim(q \lor r)$

	Statement Number	*Valid Statement*
10. P1: $(\sim p) \to q$	1	$(\sim q) \wedge [\sim(\sim r)]$
P2: $\sim[q \vee (\sim r)]$	2	$\sim q$
$\therefore\ p$	3	$(\sim q) \to [\sim(\sim p)]$
	4	$\sim(\sim p)$
	5	p
11.† P1: $p \vee (q \to r)$	1	$\sim p$
P2: $r \to s$	2	$q \to r$
P3: $s \to (\sim p)$	3	$q \to s$
P4: s		
$\therefore\ q \to s$		

The analyses in the remaining examples indicate the method of determining which valid statements lead to the desired conclusion.*

EXAMPLE 4.

If $x = 2$, then $x^2 = 4$. If $x^2 = 4$, then 4 is not an integer. But $x = 2$; therefore, 4 is not an integer.

Symbolically:

r: $x = 2$
s: $x^2 = 4$
t: 4 is an integer

P1: $r \to s$
P2: $s \to (\sim t)$
P3: r
$\therefore\ \sim t.$

Analysis: Where should one start in establishing the validity? Strange as it may seem, it is usually most helpful to work backward! The conclusion $\sim t$ is concerned directly only in P2, but P2 alone is not sufficient to establish $\sim t$ as valid; the validity of s must first be established. In turn, s follows from P1 and P3 by Rule 1. Therefore, the verification (given on the next page) begins with s.

---

*However, the remaining portion of this section can be omitted with no loss of continuity, as we shall be concerned more with analyzing valid arguments than with constructing them.

Proof:

Statement Number	Valid Statement	Reasons
1	s	P1, P3, R1
2	$\sim t$	S1, P2, R1

Argument valid.

EXAMPLE 5.

> P1: $p \rightarrow q$
> P2: $r \vee (\sim q)$
> P3: $\sim r$
> ───────
> ∴ $\sim p$

Analysis: Working backward, the conclusion concerns p, and P1 is the only premise that involves p; however P1 alone is not sufficient unless $\sim q$ can be established as valid. In turn, $\sim q$ follows from P2 and P3 by R4. Therefore, the verification begins with $\sim q$.

Proof:

Statement Number	Valid Statement	Reasons
1	$\sim q$	P2, P3, R4
2	$\sim p$	S1, P1, R3

Argument valid.

EXERCISES

1. Verify the validity of the following arguments.

   (a)† P1: p
        P2: $(p \vee q) \rightarrow r$
        ∴ r

   (b) P1: $p \rightarrow q$
        P2: $p \vee s$
        P3: $\sim s$
        ∴ q

   (c)† P1: $\sim(p \vee q)$
        P2: $s \rightarrow p$
        ∴ $\sim s$

   (d) P1: $p \rightarrow q$
        P2: $q \rightarrow r$
        P3: $r \rightarrow s$
        ∴ $p \rightarrow s$

   (e)† P1: $\sim(p \rightarrow q)$
        P2: $(p \vee q) \rightarrow s$
        ∴ s

2. Put the following arguments into symbolic form; then verify their validity.

   (a) If logic is essential for mathematics, we should study logic. Logic is essential for mathematics and mathematics is essential for a liberal education. Therefore, we should study logic.

   (b)† If we are concerned for ourselves or we are concerned for future generations, we pay heed to ecology. We do not pay heed to ecology. Therefore, we are not concerned for future generations.

   (c) If $x = 2$, then $x + 1$ is not an even number. Either $x = 2$ or $x = 5$. If $x + 4 \neq 9$, then $x \neq 5$; but $x + 4 \neq 9$. Therefore, $x + 1$ is not an even number.

10.8   INDIRECT PROOFS

Recall that the definition of a valid argument requires the implication that the conjunction of the premises implies the conclusion be always true; that is, if r denotes the conjunction of the premises and s denotes the conclusion, then $r \to s$ must always be true. The verification of each valid argument by symbolic logic up to this point was accomplished by doing precisely that—establishing that $r \to s$ is always true, either by use of truth tables or by use of the rules of inference. Such a proof is referred to as a *direct proof*.

Often times, it is more convenient to use a different approach; instead of verifying that $r \to s$ is always true, a statement equivalent to $r \to s$ is shown to be always true. Such a proof is referred to as an *indirect proof*.

This last section on logic considers two types of indirect proof often encountered in mathematics, *proof of implication* and *proof of contradiction*.

Proof of Implication

The argument

$$s \to q$$
$$\underline{s \vee (\sim p)}$$
$$\therefore\ p \to q$$

has the implication $p \to q$ as its conclusion. The proof that the argument is valid would amount to verifying the validity of $p \to q$. An examination of the rules of inference in Table 10.18 reveals that only Rule 2 is applicable in establishing that the implication would be a valid conclusion; but Rule 2 is not applicable in this case. This is unfortunate, as many of the theorems in mathematics are implications and these implications therefore are the conclusions to be established as valid. One possible approach to such a proof is to replace the implication $p \to q$ by the equivalent statement $(\sim p) \lor q$ of Pair 4 in Table 10.10; however, this is not used as often as the indirect type of proof considered here.

The indirect argument often used in this situation is based on the fact that $r \to (p \to q)$ is equivalent to $(r \land p) \to q$. Because these statements are equivalent, if $(r \land p) \to q$ is always true, then $r \to (p \to q)$ is always true also; that is, if

$$\frac{r \land p}{\therefore\ q} \qquad \text{(I)}$$

is a valid argument, then so is

$$\frac{r}{\therefore\ p \to q.} \qquad \text{(II)}$$

At first glance, Argument Forms I and II above seem too odd to be useful. However, this is not the case; remember that the statements p, q, and r can be compound statements. If r is taken to be the conjunction of the premises, the equivalence of Forms I and II indicates that if the consequent (q) is a valid conclusion from the conjunction of the premises (r) together with the antecedent (p), then $p \to q$ is a valid conclusion from the conjunction of the premises (r) alone. Rather than verify that $p \to q$ is a valid conclusion from the premises (Form II), it is sufficient to establish that q is a valid conclusion from the premises together with p (Form I).

The next several examples illustrate this method of proof.

EXAMPLE 1.

P1: $\ s \to q$
P2: $\ s \lor (\sim p)$
$\therefore\ p \to q$

Analysis: Because p is to be treated as an additional hypothesis, which we shall call an *assumption*, the first statement will indicate this assumption p is valid.

Proof:

Statement Number	Valid Statement	Reasons
1	p	Assumption
2	$\sim(\sim p)$	S1, ES
3	s	S2, P2, R4
4	q	S3, P1, R1
5	$p \to q$	S4, PI

The only change in procedure in Example 1 as compared to earlier proofs is in the first and last statements. The first statement assumes p as another premise; Statements 2, 3, and 4 establish q as a valid conclusion from the premises together with this assumption. At this point, argument Form I is completed; Statement 5 simply indicates that $p \to q$ is valid from the premises alone because of the equivalence of Form I and Form II. The notation PI (proof of implication) indicates that Statement 5 is valid from the premises because of this equivalence of the two forms.

EXAMPLE 2.

If I get a B in English or an A in Mathematics, my grade-point average will go up. I will not get an A in Mathematics. Therefore, if I get a B in English, my grade-point average will go up.

p: I get a B in English.
q: I get an A in Mathematics.
r: My grade-point average will go up.

P1: $(p \vee q) \to r$
P2: $\sim q$ _____
$\therefore p \to r$

Statement Number	Valid Statement	Reasons
1	p	Assumption
2	$p \vee q$	S1, R6
3	r	S2, P1, R1
4	$p \to r$	S3, PI

EXERCISES

Indicate the reasons for each valid statement in the indirect proof given for the following arguments.

	Statement Number	Valid Statement
1.† P1: $(p \wedge q) \to s$	1	p
P2: $\sim s$ _____	2	$\sim(p \wedge q)$
$\therefore p \to (\sim q)$	3	$(\sim p) \vee (\sim q)$
	4	$\sim(\sim p)$
	5	$\sim q$
	6	$p \to (\sim q)$

		Statement Number	*Valid Statement*
2.	P1: $(p \lor q) \to (r \lor s)$	1	p
	P2: $\sim s$	2	$p \lor q$
	$\therefore\ p \to r$	3	$r \lor s$
		4	r
		5	$p \to r$
3.†	P1: $(p \lor q) \to r$	1	q
	P2: $r \to s$	2	$p \lor q$
	$\therefore\ q \to s$	3	r
		4	s
		5	$q \to s$
4.	P1: $p \to q$	1	r
	P2: $q \to (\sim r)$	2	$\sim(\sim r)$
	$\therefore\ r \to (\sim p)$	3	$\sim q$
		4	$\sim p$
		5	$r \to (\sim p)$
5.†	P1: $p \to (q \land r)$	1	p
	P2: $(\sim r) \lor s$	2	$q \land r$
	$\therefore\ p \to s$	3	r
		4	$\sim(\sim r)$
		5	s
		6	$p \to s$

Proof by Contradiction

The second type of indirect argument is based on the pair of equivalent statements $r \to p$ and $[r \land (\sim p)] \to [q \land (\sim q)]$. Let r again denote the conjunction of the premises of an argument and let p denote the conclusion; q can denote any statement. Then, because of their equivalence, these two statements indicate that to verify that the conjunction of the premises (r) imply the conclusion (p), it is sufficient to show that the conjunction of the premises (r) together with the negation of the conclusion (p) imply that some statement (q) together with its negation ($\sim q$) are valid conclusions. In other words, if

$$\frac{\begin{array}{l} r \\ \sim p \end{array}}{\therefore\ q \land (\sim q)} \quad \text{(III)}$$

is valid, so is

$$\frac{r}{\therefore\ p.} \quad \text{(IV)}$$

EXAMPLE 3.

> P1: $p \lor q$
> P2: $\sim q$
> ───────
> $\therefore$ p

Proof:

Statement Number	Valid Statement	Reasons
1	$\sim p$	Assumption
2	q	S1, P1, R4
3	$q \land (\sim q)$	S2, P2, R7
4	p	S3, PC

The procedure in Example 3 is essentially the same as that in proving implications valid, except that in this case the negation of the conclusion is taken to be an additional premise. With Statement 3, Argument Form III is complete. That p is valid (Statement 4) from the premises alone follows from the equivalence of Forms III and IV; this is indicated by PC (proof by contradiction).

For any given argument, it is usually not obvious which will be the statement q that together with its negation, $\sim q$, will be valid. This usually requires some analysis of the argument before the proof begins. Consider another example.

EXAMPLE 4:

> P1: $r \rightarrow (s \land t)$
> P2: $\sim t$
> ───────
> $\therefore$ $\sim r$

Analysis: If we assume r as valid, where lies the contradiction? the q and $(\sim q)$? From r and P1 it follows that $s \land t$ will be valid; from $s \land t$, together with P2, both t and $\sim t$ will be valid. The proof is then structured to obtain the contradiction $t \land (\sim t)$.

Proof:

Statement Number	Valid Statement	Reasons
1	$\sim(\sim r)$	Assumption
2	r	S1, ES
3	$s \land t$	S2, P1, R1
4	t	S3, R5
5	$t \land (\sim t)$	S4, P2, R7
6	r	S5, PC

Note that this argument is the argument of Example 3 of Section 10.7.

EXAMPLE 5.

Refer to Example 2 of Section 10.7.

P1: s
P2: $t \rightarrow u$
P3: $\sim u$
∴ $(\sim t) \vee (\sim s)$

Analysis: From the negation of the conclusion, which is equivalent to $t \wedge s$, t follows as valid; from t and P2, u follows; the contradiction u and $\sim u$ then follows from u and P3.

Proof:

Statement Number	Valid Statement	Reasons
1	$\sim[(\sim t) \vee (\sim s)]$	Assumption
2	$[\sim(\sim t) \wedge \sim(\sim s)]$	S1, ES
3	$\sim(\sim t)$	S2, R5
4	t	S3, ES
5	u	S4, P2, R1
6	$u \wedge (\sim u)$	S5, P3, R7
7	$(\sim t) \vee (\sim s)$	S6, PC

Note that the first three pairs of equivalent statements in Table 10.10 give equivalent forms of the negation of compound statements. These are often helpful in proofs by contradiction, as Example 5 indicates.

EXERCISES

Give the reasons for each valid statement in the given indirect proofs of the following arguments.

	Statement Number	Valid Statement

1.† P1: $(\sim p) \rightarrow q$
    P2: $\sim q$
    ∴ p

Statement Number	Valid Statement
1	$\sim p$
2	q
3	$q \wedge (\sim q)$
4	p

(Re: Exercise 1, page 344)

2. P1: $(p \vee q) \rightarrow s$
    P2: p
    ∴ s

Statement Number	Valid Statement
1	$\sim s$
2	$\sim(p \vee q)$
3	$(\sim p) \wedge (\sim q)$
4	$\sim p$
5	$p \wedge (\sim p)$
6	s

(Re: Exercise 3, page 345)

	Statement Number	Valid Statement
3.† P1: $p \land q$	1	$\sim t$
P2: $p \to s$	2	$\sim s$
P3: $s \to t$	3	$\sim p$
$\therefore\ t$	4	p
	5	$p \land (\sim p)$
	6	t

(Re: Exercise 5, page 345)

	Statement Number	Valid Statement
4. P1: $p \lor q$	1	$\sim q$
P2: $r \to (\sim p)$	2	p
P3: r	3	$\sim(\sim p)$
$\therefore\ q$	4	$\sim r$
	5	$r \land (\sim r)$
	6	q

(Re: Exercise 6, page 345)

	Statement Number	Valid Statement
5.† P1: $(\sim p) \to q$	1	$\sim p$
P2: $\sim[q \lor (\sim r)]$	2	q
$\therefore\ p$	3	$(\sim q) \land [\sim(\sim r)]$
	4	$\sim q$
	5	$q \land (\sim q)$
	6	p

(Re: Exercise 10, page 346)

10.9 LOGIC AND MATHEMATICS— A BETWEENNESS SYSTEM

In this section, we will examine the role of logic in mathematics. For this purpose, we introduce another mathematical system, a betweenness system. A thorough logical analysis of the proofs for the theorems will be given to illustrate the logical rigor that enters into the development of mathematical systems.

First, the definition of the system is given. Then some examples and the purpose of the system are considered.

A *betweenness system* (*B*-system) is a set, *B*, of elements, together with a relation, called "between," on triples of members of *B* satisfying the axioms below.

The elements of *B* will be denoted by lower case letters, *a*, *b*, *c*, *d*, *e*, *f*, *g*, *h*, ... If *b* is between *a* and *c*, write (*abc*).

For any elements, *a*, *b*, *c*, and *d*, of *B*.

Axiom 1.   If (*abc*), then *a*, *b*, and *c* are distinct elements of *B*.

Axiom 2.   If (*abc*), then (*cba*).

Axiom 3.   If (*abc*) and (*bcd*), then (*acd*).

Axiom 4.   If (*abc*) and (*acd*), then (*bcd*).

The *B*-system is physically realized by the students in a classroom where the set *B* is the set of students and where, if *a*, *b*, and *c* denote students, *b* is between *a* and *c*—that is, (*abc*)—means that *b*'s desk is located in the same row as, and between, *a*'s desk and *c*'s desk.

Axiom 1 states that if any three elements of the set *B* have the betweenness relationship, these three elements must all be distinct; that is, no two of the three elements are the same. In the classroom situation, if *b*'s desk is in the same row as, and between, *a*'s desk and *c*'s desk—that is, (*abc*)—then *a*, *b*, and *c* are necessarily different people.

Axiom 2 states that if an element *b* is between *a* and *c*, then *b* must also be considered to be between *c* and *a*. Intuitively, this seems obvious, but properties not stated by the axioms cannot be assumed to hold!

Axiom 3 states that for any four members of *B*, *a*, *b*, *c*, and *d*, whenever *b* is between *a* and *c*, and *c* is between *b* and *d*, then it must also be that *c* is between *a* and *d*. In the classroom, if *b*'s desk is in the same row as, and between, *a*'s desk and *c*'s desk, and, in addition, if *c*'s desk is in the same row as, and between, *b*'s desk and *d*'s desk, then *c* must also be between *a* and *d*; that is, if (*abc*) and (*bcd*), then (*acd*) (Fig. 10.30). A similar analysis holds for Axiom 4.

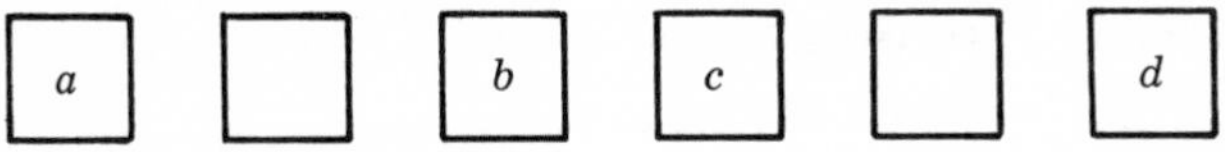

FIGURE 10.30

Other examples of the *B*-system are the colors in the spectrum, the points on a line, and a row of cars in a waiting line at a car wash where *between* is given the obvious interpretation. The set of integers, that is, the numbers . . . , −3, −2, −1, 0, 1, 2, 3, . . . , where (*abc*) means $a < b$ and $b < c$, or $a > b$ and $b > c$, is such an example.

In this last example, note that both "$a < b$ and $b < c$," and "$a > b$ and $b > c$" are required by Axiom 2. Also, equality cannot be considered here. Why?

The axioms capture the essential features of "betweenness" in these and all other such realizations of the *B*-system, and the development of the *B*-system that follows is a study of the properties of betweenness that result from these essential features, free from the influence of the nature of the elements of any one of the examples of the system. In this sense, the *B*-system is an abstract and axiomatic study of betweenness.

The theorems that follow represent part of this study. Each of these theorems could arise, for example, from examining the arrangement of students in a classroom or points on a line, and then conjecturing that order properties that hold in these arrangements hold in each *B*-system. That is, possible theorems can arise from consideration of one or two examples. As always, the proofs must be established in the general framework given by the axioms. Theorems can also arise simply by considering what additional relationships on the members of *B* are demanded by the axioms, apart from any one example.

Consider now four points, *e*, *f*, *g*, and *h*, located on a line as indicated in Figure 10.31. In the betweenness system consisting of the points on the line, we have (*efg*) and (*fgh*). Because the conditional part of Axiom 3 is satisfied for these four points, it follows from Axiom 3 that (*egh*) must also hold; this is geometrically verified in Figure 10.31.

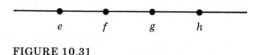

$$
\begin{array}{cccc}
e & f & g & h
\end{array}
$$

FIGURE 10.31

Note also that (*efh*), which is not required by any one of the axioms, also holds on this line. Possibly these relationships hold in every *B*-system; that is, if (*efg*) and (*fgh*), then (*efh*). Theorem 1 states that they do. The proof of the theorem is the verification that the statement of the theorem is a logical consequence of the axioms.

Theorem 1. For any four members of *B*, *e*, *f*, *g*, and *h*, if (*efg*) and (*fgh*), then (*efh*).

Proof:

Statement Number	Valid Statement	Reasons
1	(*efg*) and (*fgh*)	Assumption
2	(*fgh*)	S1, R5
3	(*hgf*)	S2, A2, R1 (see Comment 1)
4	(*efg*)	S1, R5
5	(*gfe*)	S4, A2, R1 (see Comment 2)
6	(*hgf*) and (*gfe*)	S3, S5, R7
7	(*hfe*)	S6, A3, R1 (see Comment 3)
8	(*efh*)	S7, A2, R1 (see Comment 4)
9	[(*efg*) and (*fgh*)] → (*efh*)	S8, PI

Comments:

1. For the elements f, g, and h, the statement of Axiom 2 becomes "$(fgh) \rightarrow (hgf)$." The argument by which Statement 3 is valid is therefore

   $\quad$ $(fgh) \rightarrow (hgf)$ $\qquad$ (A2)

   $\quad$ $\underline{(fgh)}$ $\qquad\qquad\quad$ (S2)

   $\therefore$ (hgf),

   which is in the form of Rule 1.

2. For the elements e, f, and g, the statement of Axiom 2 becomes "$(efg) \rightarrow (gfe)$." The argument by which Statement 5 is valid is therefore

   $\quad$ $(efg) \rightarrow (gfe)$ $\qquad$ (A2)

   $\quad$ $\underline{(efg)}$ $\qquad\qquad\quad$ (S4)

   $\therefore$ (gfe),

   which is again in the form of Rule 1.

3. For the elements h, g, f, and e, the statement of Axiom 3 becomes "$[(hgf)$ and $(gfe)] \rightarrow (hfe)$." The application of Rule 1 becomes

   $\quad$ $[(hgf)$ and $(gfe)] \rightarrow (hfe)$ $\qquad$ (A3)

   $\quad$ $\underline{(hgf)}$ and (gfe) $\qquad\qquad\quad\;$ (S6)

   $\therefore$ (hfe).

4. Rule 1 is again applied as above with the statement of Axiom 2 being "$(hfe) \rightarrow (efh)$."

Again, consider three points as in Figure 10.32, with a fourth point, h, located as in Figure 10.33 or as in Figure 10.34.

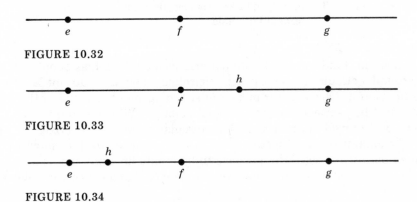

FIGURE 10.32

FIGURE 10.33

FIGURE 10.34

In either case, it is geometrically evident that (*fgh*) cannot hold. This fact leads to the conjecture that if (*efg*) and (*ehg*) hold in any *B*-system, then (*fgh*) cannot hold. The proof of Theorem 2 verifies that the conjecture is in fact valid.

Theorem 2.   For any four members of *B*, *e*, *f*, *g*, and *h*, if (*efg*) and (*ehg*), then (*fgh*) cannot hold.

Proof:

Statement Number	Valid Statement	Reasons
1	~{[(*efg*) and (*ehg*)] → [~(*fgh*)]}	Assumption
2	[(*efg*) and (*ehg*)] and ~[~(*fgh*)]	S1, ES
3	(*efg*) and (*ehg*)	S2, R5
4	(*efg*)	S3, R5
5	~[~(*fgh*)]	S2, R5
6	(*fgh*)	S5, ES
7	(*efg*) and (*fgh*)	S4, S6, R7
8	(*egh*)	S7, A3, R1
9	(*ehg*)	S3, R5
10	(*ehg*) and (*egh*)	S8, S9, R7
11	(*hgh*)	S10, A4, R1
12	~A1	S11
13	A1 and ~A1	S12, A1, R7
14	[(*efg*) and (*ehg*)] → [~(*fgh*)]	S13, PC

The conjecture that Theorem 3 might be valid could also arise by considering the relationships between points on a line in the same way as Theorems 1 and 2. The statement of Theorem 3 follows; its proof is given in the exercises.

Theorem 3.   If *e*, *f*, and *g* are members of *B*, (*efg*) and (*fge*) cannot both hold.

Theorem 3 indicates that circular arrangements, such as people sitting around a circular table or pearls encircling a necklace, do not satisfy the axioms of a *B*-system, for in a circular arrangement any one of the elements must be between every two other elements. Which axiom would have to be removed in order to include circular arrangements (Exercise 4)?

The axioms of the *B*-system allow for physical realizations in which the members of the set *B* do not have to be in one line, but can be in several lines, as the students in a classroom or the cars in a parking lot. What axiom should be added to the set of axioms of the *B*-system to allow only instances where the elements are in one line (Exercise 5)?

Note the presence of the ingredients of any mathematical system in the *B*-system. There is only one defined term, the *B*-system itself; the undefined terms are the *elements* of *B* and *relation of betweenness* on triples of members of *B*. Many nontechnical terms appear. The axioms and theorems are easy to identify. More theorems can be deduced, but our concern here is not the full development of the system, but rather a glimpse into its development and the role of logic in mathematics.

While the formal aspect of mathematics, the verification that theorems are consequences of the axioms, is an exercise of *deductive* logic, note the following two ways in which *inductive* logic enters into mathematics:

1. The axioms of the *B*-system were determined by presenting the essential features of betweenness as realized in the various examples given.
2. Suggestions for possible theorems came from examining one or more examples.

This determination of axioms and possible theorems from the analysis of various examples represents a movement from particular instances to general conclusions. Such a movement is an exercise of inductive logic, which therefore has its role to play in mathematics as well.

EXERCISES

1. Give two examples of a *B*-system other than those given in the text.

2.† The proof of Theorem 3 is given below. Give the reasons why the various statements are valid.

Theorem 3.   For any three members of *B*, *e*, *f*, and *g*, (*efg*) and (*fge*) cannot both hold.

Proof:

Statement Number	Valid Statement	Reasons
1	$\sim\{\sim[(efg) \text{ and } (fge)]\}$	
2	(*efg*) and (*fge*)	
3	(*ege*)	
4	$\sim$A1	
5	A1 and $\sim$A1	
6	$\sim[(efg) \text{ and } (fge)]$	

3. If e, f, g, h, and i are points on a line as indicated in Figure 10.35, then (efg), (fgh), and (ghi) all hold.

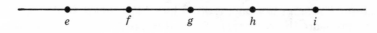

e   f   g   h   i

FIGURE 10.35

The locations of these five points suggest other possible theorems in the B-system.

(a) Determine two such possible theorems by determining two conclusions for:

"If e, f, g, h, and i are members of B such that (efg), (fgh), and (ghi), then . . . ."

(b) Choose one of the completed statements in part (a), and determine whether in fact it is a theorem by attempting to construct a proof.

*4.† Which axiom should be eliminated to allow for circular arrangements in the B-system? (Refer to page 358.)

*5. Which axiom should be added to allow only arrangements in one line in the B-system? (Refer to page 359.)

*6.† What changes should be made to the set of axioms of the B-system to allow for circular arrangements, but to exclude arrangements in more than one circle or line?

*7. In the example of a B-system consisting of the points on a line, note that between every two points on the line there is a third point. This observation leads to the conjecture, "For every two members, a and b, of B there is a third member, c, of B such that (acb)." Why can we say that this is not a valid conclusion of the axioms of the B-system prior to any proof? (*Hint:* Consider other examples.)

10.10 SUMMARY—THE AUTHORITY OF MATHEMATICS

Symbolic logic provides a means of analyzing deductive reasoning based only on the form of the argument, independent of the content of the statements. The central idea is the determination of the validity of an argument. If r denotes the conjunction of the premises and s the conclusion, then the argument $r \rightarrow s$ is valid if and only if $r \rightarrow s$ is always true. An alternative to verifying directly that $r \rightarrow s$ is always true is to verify that a statement equivalent to $r \rightarrow s$ is always true. The latter is referred to as an indirect proof.

In the previous section, it was indicated that the colors of the spectrum form a B-system, where betweenness is determined by the location of the colors in the spectrum. Hence, the B-system serves as a mathematical model of the spectrum with respect to the location of the colors.

Suppose that it is known only that yellow (f) is between red (e) and violet (g) and that green (h) is between red (e) and violet (g), and the question is whether violet (g) is between yellow (f) and green (h). Theorem 2 of the B-system, which states that if (efg) and (ehg), then (fgh) cannot hold, indicates that it is not. How can we be certain that this answer is correct, that violet is not between yellow and green?

In general, great authority is usually associated with applications of mathematics in the sense that whatever is mathematically determined must be correct. What is the source of this authority, assuming that no errors in the use of mathematics occurred? It is here that "what is *true*" and "what is *valid*" join together.

Recall that *validity* is concerned with logical consequences and *truth* is concerned with correspondence with reality. Recall also that theorems are logical consequences of the axioms of a mathematical system. Hence, if r denotes the conjunction of the axioms and s denotes the theorem, $r \rightarrow s$ is a true statement for all combinations of truth values of r and s.

Suppose now that an appropriate mathematical model is chosen to solve a problem. Then each axiom is true in that each axiom is physically realized; each axiom corresponds to reality. It follows that the r of $r \rightarrow s$ is true. Can s be false; that is, can s not correspond to reality? No, s must also be true; otherwise $r \rightarrow s$ would then be false! It is here that the authority of mathematics in applications has its foundation—if each axiom is true in a physical situation, then each theorem must also be true.

In our illustration, the mathematical description of the spectrum was the B-system, and a theorem of this system was used to answer a question about the spectrum. In this case, each axiom of the B-system is true in the physical reality of the spectrum. It therefore follows that Theorem 2, which was used to answer the question, must also be true of the spectrum.

appendix a

conversion to the metric system

The need for, and natural knowledge of, a system of weights and measures was beautifully expressed by John Quincy Adams, then Secretary of State, in a report to Congress in 1821:

> Weights and measures may be ranked among the necessaries of life to every individual of human society. They enter into the economical arrangements and daily concerns of every family. They are necessary to every occupation of human industry; to the distribution and security of every species of property; to every transaction of trade and commerce; to the labors of the husbandman; to the ingenuity of the artificer; to the studies of the philosopher; to the researches of the antiquarian, to the navigation of the mariner, and the marches of the soldier; to all the exchanges of peace, and all the operations of war. The knowledge of them, as in established use, is among the first elements of education, and is often learnt by those who learn nothing else, not even to read and write. This knowledge is rivetted in the memory by the habitual application of it to the employments of men throughout life.

The English system of weights and measures, with foot, pound, quart, and so forth as the standard units, is the most common system "in established use" in the United States today. By the eighteenth century, this system, the roots of which go back to Roman civilization and beyond, had become the most standardized system of western civilization, due principally to the English. Through close political and economic association with England, the system spread to the American colonies as well as other parts of the world.

At the end of the eighteenth century, the French Academy of Science, at the request of the National Assembly of France, appointed a commission to establish a new system of weights and measures. The result was the metric system.

The basic unit of measurement established by the commission was the *meter*, a unit of length equal to one ten-millionth of the distance from the north pole to the equator measured along a meridian of the earth near Dunkirk, France.

362

The basic unit of volume was the *liter*, defined to be the volume of a cube one tenth of a meter on each of its sides. The basic unit of mass (weight) was the *gram*, defined to be the mass of a cube of water when the water is at its maximum density and the cube is one hundredth of a meter on each of its sides. These three are still basic units of measurement in the metric system, but their definitions have changed, as indicated below. Other units of measurement were obtained by multiplying or dividing the three basic units, meter, liter, and gram, by powers of ten, yielding units such as the centimeter (1/100 meter) and the kilogram (1000 grams).

The metric system was made compulsory in France in 1840 and gradually was adopted by other nations. It is the most common system in practically all the industrialized world today, with the singular exception of the United States.

Control of the metric system at the present time is in the hands of the General Conference of Weights and Measures, which was established by an international treaty of 1875 as a permanent agency to periodically review and ratify changes to the system. (The United States was one of the signers of the treaty.) The last extensive revision adopted by the conference was in 1960; the metric system as adopted at that time is referred to as the International System of Units (SI).

Recognizing the need for uniformity among the states, the framers of the Constitution gave to the Congress the power to establish a system of weights and measures. At the present time, The National Bureau of Standards, a branch of the Department of Commerce, is the agency responsible for maintaining uniform standards for the country.

In 1866, when Congress legalized its use, the metric system became the only legal measurement system in the United States. The English system, while traditional, never had the official approval of Congress. Through the years, various efforts were made to change the United States to predominant use of the metric system; these efforts did not go unopposed. Finally, on December 23, 1975, President Gerald Ford signed into law a bill passed by Congress committing the United States to a metrication program.

A meaningful adoption of metric by the layman requires two things:

1. A feeling for the size of metric units.
2. An ability to convert from the English system to metric, and vice versa.

Tables A.1 and A.2, adapted from The National Bureau of Standards Letter Circular 1051, July 1973, are designed to enable the reader to achieve these two objectives. Table A.1 contains most of the terms required for everyday usage of the metric system.

Note that variations of the basic units are obtained by a change in the location of the decimal point. For example, 2.0 meters is the same as 0.002 kilometers or 200.0 centimeters. For this reason, the metric system is

TABLE A.1　Metric System Units (International System of Units—SI)

Basic Units	Technical Definition
METER: a little longer than a yard (about 1.1 yards)	1 650 763.73* wavelengths in a vacuum of the orange-red line in the spectrum of krypton-86
LITER: a little larger than a quart (about 1.06 quarts)	fluid volume of 0.001 cubic meter
GRAM: about the weight of a paper clip	0.001 of the mass of a cylinder of platinum-iridium alloy kept by the International Bureau of Weights and Measures in Paris

Common Prefixes (to be used with basic units)

Kilo:	one thousand (1,000)	For example:
Hecto:	one hundred (100) ⎫ seldom	1 millimeter = 0.001 meter
Deca:	ten (10) ⎬ used	1 centimeter = 0.01 meter
Deci:	one-tenth (0.1) ⎭	1 kilometer = 1,000 meters
Centi:	one-hundredth (0.01)	
Milli:	one-thousandth (0.001)	

Commonly Used Units

Millimeter:	0.001 meter	diameter of a paper clip wire
Centimeter:	0.01 meter	width of a paper clip (about 0.4 inch)
Kilometer:	1,000 meters	somewhat further than $\frac{1}{2}$ mile (about 0.6 mile)
Kilogram:	1,000 grams	a little more than 2 pounds (about 2.2 pounds)
Milliliter:	0.001 liter	five of them make a teaspoon

Other Useful Units

Hectare:	10,000 square meters (about $2\frac{1}{2}$ acres)
Metric ton:	1,000 kilograms (about one ton)

Temperature (degrees Celsius are used)

°C	-40	-20	0	20	37	60	80	100
°F	-40		0	32	80　98.6		160	212
				water freezes	body temperature			water boils

Comparison of Centimeters and Inches

cm　1　2　3　4　5　6　7　8　9　10

inches　1　2　3　4

*Spaces are used instead of commas to separate digits; however four-digit numbers are normally not separated unless used in **columns**.

book

TABLE A.2   Metric Conversion Factors

Approximate *Conversions to Metric Measures*[1]				
Symbol	*When You Know*	*Multiply By*	*To Find*	*Symbol*
		LENGTH		
in.	inches	2.54	centimeters	cm
ft	feet	30.48	centimeters	cm
yd	yards	0.91	meters	m
mi	miles	1.61	kilometers	km
		AREA		
$in.^2$	square inches	6.45	square centimeters	cm^2
ft^2	square feet	0.09	square meters	m^2
yd^2	square yards	0.84	square meters	m^2
mi^2	square miles	2.59	square kilometers	km^2
	acres	0.40	hectares	ha
		MASS (weight)		
oz	ounces	28.35	grams	g
lb	pounds	0.45	kilograms	kg
	short tons (2,000 lb)	0.91	metric tons	t
		VOLUME		
tsp	teaspoons	5	milliliters	ml
Tbsp	tablespoons	15	milliliters	ml
fl oz	fluid ounces	30	milliliters	ml
c	cups	0.24	liters	ℓ
pt	pints	0.47	liters	ℓ
qt	quart	0.95	liters	ℓ
gal.	gallons	3.79	liters	ℓ
ft^3	cubic feet	0.03	cubic meters	m^3
yd^3	cubic yards	0.76	cubic meters	m^3
		TEMPERATURE (exact)		
°F	Fahrenheit temperature	5/9 (after subtractting 32)	Celsius temperature	°C

1. For exact conversions and more detailed tables, see NBS Misc. Publ. 286, *Units of Weights and Measures.*

TABLE A.2 (continued)

	Approximate *Conversions from Metric Measures*			
Symbol	*When You Know*	*Multiply By*	*To Find*	*Symbol*
		LENGTH		
mm	millimeters	0.04	inches	in.
cm	centimeters	0.39	inches	in.
m	meters	3.28	feet	ft
m	meters	1.09	yards	yd
km	kilometers	0.62	miles	mi
		AREA		
cm^2	square centimeters	0.16	square inches	$in.^2$
m^2	square meters	1.20	square yards	yd^2
km^2	square kilometers	0.39	square miles	mi^2
ha	hectares ($10,000\ m^2$)	2.47	acres	
		MASS (weight)		
g	grams	0.035	ounces	oz
kg	kilograms	2.2	pounds	lb
t	metric tons	1.1	short tons	
		VOLUME		
ml	milliliters	0.03	fluid ounces	fl oz
ℓ	liters	2.11	pints	pt
ℓ	liters	1.06	quarts	qt
ℓ	liters	0.26	gallons	gal.
m^3	cubic meters	35.31	cubic feet	ft^3
m^3	cubic meters	1.31	cubic yards	yd^3
		TEMPERATURE (exact)		
°C	Celsius temperature	9/5 (then add 32)	Fahrenheit temperature	°F

referred to as a *decimal* system. Furthermore, shifting the decimal is a simpler procedure than, say, dividing inches by 12 to obtain feet or multiplying pounds by 16 to obtain ounces. Because of this simplicity, the metric system is particularly suited to scientific work.

The metric units given in Tables A.1 and A.2 are those that one might encounter in daily usage. The complete metric (SI) system involves other technical units for engineering and scientific work; for example, the *pascal* for measuring pressure and the *lumen* for measuring light flux.

Through the use of the conversion tables, the process of changing English system units into metric, and vice versa, becomes a simple arithmetic problem, as illustrated by the following examples.

EXAMPLE 1.

The distance from Toledo to Detroit is 50 miles (mi). How many kilometers (km)?

According to Table A.2, to convert miles into kilometers, multiply the number of miles by a factor of 1.61. The distance is therefore 50 × 1.61 = 80.5 km.

EXAMPLE 2.

A package contains $\frac{1}{2}$ kilogram (kg) of cookies. How many pounds (lb)? How many ounces (oz)?

According to Table A.2, to convert kilograms into pounds, multiply the number of kilograms by 2.2. There are therefore $\frac{1}{2}$ × 2.2 = 1.1 lb of cookies.

The table does not give a conversion factor for kilograms into ounces; however, because each pound contains 16 ounces, there are 1.1 × 16 = 17.6 oz of cookies.

EXAMPLE 3.

It costs $92.00 to carpet a room measuring 12 square yards (yd^2) with a particular type of carpet. What is the cost per square meter (m^2)?

Because the room measures 12 × 0.84 = 10.08 square meters, the cost per square meter is $92 ÷ 10.08, or approximately $9.13. (Note that this is the same price as $92 ÷ 12, or approximately $7.67 per square yard.)

EXAMPLE 4.

Milk that sells at 45¢ per liter (ℓ) costs how much per quart (qt)?

Because one liter is equal to 1.06 quarts, the cost per quart is 45 ÷ 1.06, or approximately 42¢.

EXERCISES

1.† An automobile gasoline tank holds 20 gallons (gal.). How many liters (ℓ)?

2. A $\frac{1}{4}$ inch (in.) nut requires a wrench of how many centimeters (cm)?

3.† A 12 ounce (oz) box of cereal contains how many grams (g)?

4. If a bottle of beer contains 12 fluid ounces (fl oz), how many liters (ℓ) does it contain?

5.† A 50 yard (yd) dash is a dash of how many meters (m)?

6. When the temperature is 78 degrees Fahrenheit (°F), what is the temperature in degrees Celsius (°C)?

7. Approximate the number of meters (m) from your dormatory (or parking lot) to your first class of the week.

8. Determine your height in centimeters (cm) and your weight in both kilograms (kg) and grams (g).

9.† If the distance from Toledo to Cleveland is 160 kilometers (km), what is the distance in miles (mi)?

10. The daily dosage of a particular antibiotic is 3 grams (g). What is the daily dosage in ounces (oz)?

11.† A 60 hectare (ha) park contains how many acres?

12. A 25 ounce (oz) jar of peanut butter costs 65 cents. What is the price per gram (g)?

13.† If a 5 pound (lb) bag of sugar costs $1.20, what is the price per kilogram (kg)?

14. Bob runs 1,500 meters (m) in 4 minutes. How many yards (yd) does he run per minute?

15.† Which is more economical to buy, a gallon (gal.) of paint for $5.25 or 4 liters (ℓ) of the same paint for the same price?

16. Which is more economical to buy, a 16 ounce (oz) can of tomatoes for 33¢ or a 450 gram (g) can of the same tomatoes for 37¢?

17.† The directions for mixing a particular type of patching plaster calls for 1 cup (c) of water for each half kilogram (kg) of plaster. How much water should be mixed with one pound (lb) of the plaster?

A.1 AREA AND VOLUME

This section presents a brief treatment of the calculation of the area and volume of various geometric figures that occur in ordinary usage of mathematics. The formulas for these areas and volumes are given in Tables A.3 and A.4.

TABLE A.3

Area (in square units)	
Rectangle: ℓw	(length times width)
Circle: πr^2	(π times the radius squared)
Triangle: $\frac{1}{2} bh$	($\frac{1}{2}$ times the base times the height)

TABLE A.4

Volume (in cubic units)	
Rectangular Solid: ℓwh (a box)	(length times width times height)
Sphere: $\frac{4}{3} \pi r^3$ (a ball)	($4\pi/3$ times the radius cubed)
Right Circular Cylinder: $\pi r^2 h$ (a can)	(π times the radius squared times height)

EXAMPLE 1.

Determine the area of the triangle in Figure A.1.

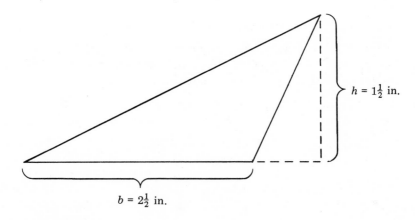

$h = 1\frac{1}{2}$ in.

$b = 2\frac{1}{2}$ in.

FIGURE A.1

The area of a triangle is $\frac{1}{2} bh$, so the area of this particular triangle is

$(\frac{1}{2})(\frac{5}{2})(\frac{3}{2}) = \frac{15}{8} = 1\frac{7}{8}$ square inches.

EXAMPLE 2.

A quick delivery service will accept only packages which are less than 27 cubic feet in volume. Mary Lou wants a carton of dishes delivered that measures 30 inches by 30 inches by 3 feet. Will the delivery service handle the package?

The volume of a carton is ℓwh, so the volume of the carton containing the dishes is

$$\left(\tfrac{5}{2}\right) \times \left(\tfrac{5}{2}\right) \times 3 = \tfrac{75}{4} = 18\tfrac{3}{4} \text{ cubic feet.}$$

The carton is within the acceptable limits.

EXAMPLE 3.

The spherical storage tank has a radius of 6 feet. How many cubic meters does it contain?

The volume of a sphere is $\tfrac{4}{3}\pi r^3$ and π is approximately 3.14, so the tank contains approximately

$$\left(\tfrac{4}{3}\right)(3.14)6^3 = 904.32 \text{ cubic feet.}$$

By the metric conversion tables, this would be

$$904.32 \times 0.03 = 27.1296 \text{ cubic meters.}$$

EXAMPLE 4.

How many gallons does the tank in Example 3 contain?

Table A.2 gives the conversion factor only from liters into gallons. Because a liter is, by definition, 0.001 cubic meters, each cubic meter is 1000 liters. The tank contains 27 129.6 liters, or

$$27\ 129.6 \times 0.26 = 7053.696 \text{ gallons.}$$

EXERCISES

1.† Determine the volume, in cubic yards, of a sphere with a radius of 3.5 meters.

2. The rules for entering a sailing regatta require that each boat have a single sail of no more than 25 square feet. Sally's boat has a triangular sail that is 10 feet high and measures 6 feet at the base. Can Sally enter the regatta?

3.† The specifications on a can of paint indicate that one quart covers 8 square meters. Dan intends to paint both sides of a circular disk, radius 8 feet, with this paint. How many quarts should he buy?

*4. A can of tomato juice has a radius of 2 inches and is 8 inches tall. How many liters does it contain?

5.† A lawn chemical service charges $15.00 for servicing a lawn less than 5,000 square feet and $19.00 for any larger lawn. Jim's front yard measures 40 by 60 feet and his back yard 40 by 40 feet. Will he have to pay $15 or $19 for the lawn service? (His side yard is paved.)

6. A cake recipe indicates that the batter should be placed in two cake pans with a 12-inch diameter and a depth of $1\frac{1}{4}$ inches. Eve has only a rectangular cake pan 9 by 12 by 2 inches deep. Can the rectangular pan contain the cake?

appendix b
answers to
selected exercises

ANSWERS FOR CHAPTER 1

Section 1.1 (page 5)

1. (a)

Node	Order
A	1
B	3
C	3
D	1

(c)

Node	Order
A	3
B	2
C	3
D	2
E	3
F	3

2. (a)

Node	Type
A	Odd
B	Odd
C	Odd
D	Odd

Sum is 8.

(c)

Node	Type
A	Odd
B	Even
C	Odd
D	Even
E	Odd
F	Odd

Sum is 16.

3. Graphs (a) and (c) of Figure 1.8 cannot be drawn in the manner indicated.

4. Two of many possible graphs are shown in Figure B.1:

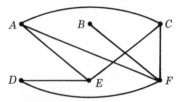

A, C, and E are all acquainted.   A, B, and C are complete strangers.

FIGURE B.1

Section 1.1 (page 10)

1. (a)   No Euler path (Theorem 4).
    (c)   One of many possible Euler paths is *AaCCcEEdCCbBBjEEhDDgEEiAAeDDfB* (Theorem 6).

3. (a)   There can be no Euler circuit; one of the many possible circuits is *AaBBbCCcDDdEEeFFfGGgHHhA*.
    (c)   There can be no Euler circuit; one of the many possible circuits is *AeDDfBBjEEdCCaA*.

5. (a)   The graph (Fig. B.3) of the puzzle (Fig. B.2) has more than two odd nodes.

    (b)

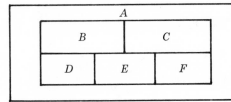

FIGURE B.2

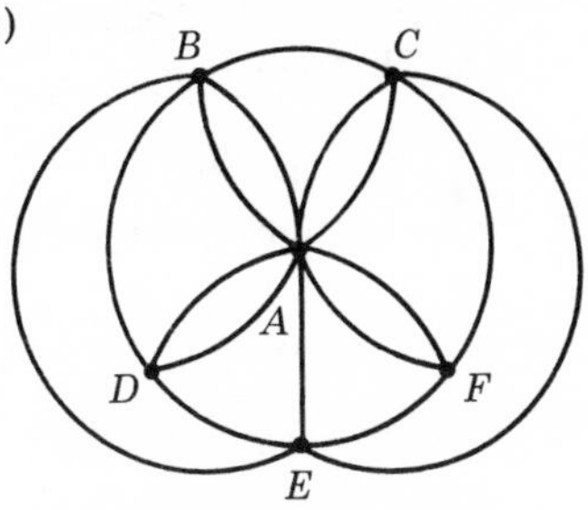

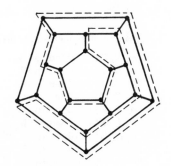

FIGURE B.3

Section 1.2 (page 15)

1. (a)   One possible Euler circuit: *ADCBABDCA*. One possible Hamilton circuit: *DBACD*.
    (c)   No Euler circuit possible. One possible Hamilton circuit: *FABECDF*.

3. One possible solution is indicated by the dotted line in Figure B.4.

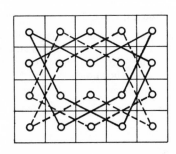

FIGURE B.4

5. (a)   The orders for the corresponding nodes are indicated in Figure B.5. Because there are odd nodes, no Euler circuit is possible.
    (c)   Minimum number is two (Fig. B.6):

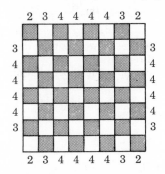

FIGURE B.5

FIGURE B.6

Section 1.3 (page 22)

1. (a)   Bipartite.   (c)   Not bipartite.
2. (a)   Theorem 7 guarantees that there is no Hamilton circuit for the graph in Exercise 1(a); it tells us nothing about the graph in Exercise 1(c).
    (b)   The graph in Exercise 1(c) does have a Hamilton circuit.
3. Because the knight always alternates between black and red squares when it moves, the graph representing all of the possible knight moves is a bipartite graph. When one corner is removed, the dividing sets can no longer have the same number of nodes. Consequently, a Hamilton circuit is no longer possible by Theorem 7.

Section 1.3 (page 26)

1. One possible match is given by Figure B.7.

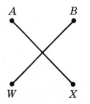

FIGURE B.7

3. One possible solution is given by Figure B.8.

$\{A, B, C\}$ is joined to $\{W, X, V\}$
$\{A, B. D\}$ is joined to $\{W, X, Y\}$
$\{A, B, E\}$ is joined to $\{W, X, Y\}$
$\{A, C, D\}$ is joined to $\{V, X, Z \}$
$\{A, C, E\}$ is joined to $\{V, X, Z \}$
$\{A, D, E\}$ is joined to $\{W, X, Y\}$
$\{B, C, D\}$ is joined to $\{V, W, Y\}$
$\{B, C, E \}$ is joined to $\{V, W, Y\}$
$\{B, D, E\}$ is joined to $\{W, X, Y\}$
$\{C, D, E\}$ is joined to $\{X, Y, Z \}$

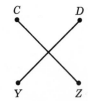

FIGURE B.8

However, $\{A, B, D, E\}$ is joined only to $\{W, X, Y\}$.

5. $\{C_1, C_2, C_3\}$ is joined only to $\{M_3, M4\}$. Consequently, the committees cannot be matched to faculty members.

Section 1.4 (page 32)

2. (a)   In a planar graph, the number of branches is two less than the sum of the number of nodes and the number of regions.
    (b)   No. In the graph of Figure B.9, for example, the above relationship does not hold.

FIGURE B.9

4. Inductive

ANSWERS FOR CHAPTER 2

Section 2.1 (page 44)

1. (a) $A \cup B = \{-6, -4, -2, 0, 1, 2, 3, 4, 5, 6\}$
    (b) $A \cap B = \{2, 4, 6\}$
    (c) $(A \cap B) \cup C = \{-6, -3, 0, 2, 3, 4, 6\}$
    (d) $A' = \{-5, -3, -1, 1, 3, 5\}$
    (e) $C' = \{-5, -4, -2, -1, 1, 2, 4, 5\}$
    (f) $A' \cup C' = \{-5, -4, -3, -2, -1, 1, 2, 3, 4, 5\}$
    (g) $(A \cup C)' = \{-5, -1, 1, 5\}$

3. (a) $2 \in A$ (b) $D \subset A$ (c) $A \not\subset B$ (d) $\{3\} \subset C$
    (e) $\emptyset \subset C$ (f) $U \not\subset B$ (g) $3 \notin D$

5. (a) B': the shaded area in Figure B.10.
    (c) $(A \cap B)'$: the shaded area in Figure B.11.
    (e) $A \cap B'$: the shaded area in Figure B.12.

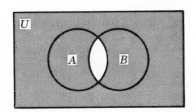

FIGURE B.10.

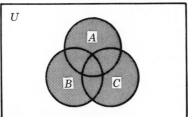

FIGURE B.11.

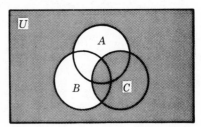

FIGURE B.12.

6. (a) $(A \cup B) \cup C$: the shaded area in Figure B.13.
    (c) $(A \cup B)' \cup C$: the shaded area in Figure B.14.

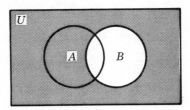

FIGURE B.13

FIGURE B.14.

Section 2.2 (page 47)

1. (a) $\{1, 3, 5\}$ (b) $\{1, 2, 3\}$ (c) $\{1, 2, 3, 5, 6\}$

3. (a) $\{(HHT), (HTH), (THH)\}$ (b) $\{(HHH), (HHT), (HTH), (THH)\}$
    (c) $\{(TTT)\}$

5. (a) $S = \{(1,1), (1,2), (1,3), (1,4), (1,5), (1,6), (2,1), (2,2), (2,3),$
$(2,4), (2,5), (2,6), (3,1), (3,2), (3,3), (3,4), (3,5), (3,6),$
$(4,1), (4,2), (4,3), (4,4), (4,5), (4,6), (5,1), (5,2), (5,3),$
$(5,4), (5,5), (5,6), (6,1), (6,2), (6,3), (6,4), (6,5), (6,6)\}$

    (b) (i) $\{(1,4), (4,1), (2,3), (3,2)\}$
    (ii) $\{(1,1), (1,3), (3,1), (1,2), (2,1), (2,2)\}$
    (iii) $\{(1,2), (2,1), (1,4), (4,1), (2,3), (3,2)\}$

Section 2.2 (page 49)

1. (a) $E_1 \cap E_2 ; \{(135), (235), (345)\}$
2. (b) $E_1 \cup E_2 ; \{(1,1), (1,2), (1,3), (1,4), (1,5), (1,6), (2,2),$
$(3,1), (5,1), (2,4), (4,2), (3,3), (2,6), (6,2),$
$(3,5), (5,3), (4,4), (6,4), (4,6), (5,5), (6,6)\}$

3. (c) $E_2' ; \{(1,2), (1,3), (1,4), (1,5), (1,6), (2,1), (2,3), (2,4), (2,5),$
$(2,6), (3,1), (3,2), (3,4), (3,5), (3,6), (5,1), (5,2), (5,3),$
$(5,4), (5,6), (6,1), (6,2), (6,3), (6,4), (6,5)\}$

Section 2.3 (page 53)

1. (a) $\frac{3}{6} = \frac{1}{2}$ (b) $\frac{3}{6} = \frac{1}{2}$ (c) $\frac{5}{6}$
3. (a) $\frac{18}{38} = \frac{9}{19}$ (b) $\frac{4}{38} = \frac{2}{19}$ (c) $\frac{12}{38} = \frac{6}{19}$
    (d) All four events contain the same number of elements, namely 18; all have
         the probability $\frac{18}{38} = \frac{9}{19}$.

5. $\frac{15}{36} = \frac{5}{12}$

7. (a) 0.7   (b) 0.9   (c) 0.6   (d) 1
    (e) 0   (f) 1   (g) 0.8

9. (a) 0.8   (b) 0.1   (c) 0.4   (d) 0.9

Section 2.3 (page 59)

1. $P(E') = P(b) + P(c) = 0.4$ $P(E') = 1 - P(E) = 0.4$

3. (a) $P(S) = P(\alpha) + P(\beta) + P(\gamma) = 0.9$; but $P(S)$ must always be 1.
    (c) $P(E) + P(E') = 1.1$; but $P(E) + P(E')$ must always be 1.
    (e) Because $E_3 = E_1 \cup E_2$, and $E_1 \cap E_2 = \emptyset$, $P(E_3)$ should be the same as
         $P(E_1) + P(E_2)$.

4. $P(E_1 \cup E_2) = P(E_1) + P(E_2) - P(E_1 \cap E_2)$

5. (b) $P(E_1 \cup E_2) = 0.6$ (d) $P(E_1 \cup E_1') = 1$

6. (a) $P(E_1) = \frac{15}{36} = \frac{5}{12}$ $P(E_2) = \frac{5}{12}$ $P(E_3) = \frac{6}{36} = \frac{1}{6}$
    (c) The event that the house wins is $E_2 \cup E_3$;
         $P(E_2 \cup E_3) = P(E_2) + P(E_3) = \frac{5}{12} + \frac{1}{6} = \frac{7}{12}$ because $E_2 \cap E_3 = \emptyset$.

Section 2.4 (page 63)

1. (a) $0.35/0.65 = 7/13$ (b) 13:7

3. (a) 9:10   (b) 10/9 or about $1.11

5. \$3.50

7. (a) 1:7   (b) Yes

ANSWERS FOR CHAPTER 3

Section 3.2 (page 73)

1. (a)   Commutative Law of Addition
    (b)   Associative Law of Addition
    (e)   Associative Law of Addition
    (g)   Commutative Law of Addition

2. (a) $a + (b + c) =$ by the Associative Law of Addition
          $(a + b) + c =$ by the Commutative Law of Addition
          $(b + a) + c$
    (c) $(3 + a) + 5 =$ by the Commutative Law of Addition
          $(a + 3) + 5 =$ by the Associative Law of Addition
          $a + (3 + 5)$

Section 3.2 (page 74)

1. (a)   Commutative Law of Multiplication
    (b)   Associative Law of Multiplication
    (c)   Distributive Law
    (f)   Commutative Law of Multiplication

2. (a) $4 \times (a \times b) =$ by the Associative Law of Multiplication
          $(4 \times a) \times b =$ by the Commutative Law of Multiplication
          $(a \times 4) \times b$
    (c) $(c \times 12) \times a =$ by the Associative Law of Multiplication
          $c \times (12 \times a) =$ by the Commutative Law of Multiplication
          $c \times (a \times 12) =$ by the Associative Law of Multiplication
          $(c \times a) \times 12$
    (e) $a \times [(b \times (c + 2)] =$ by the Distributive Law
          $a \times [(b \times c) + (b \times 2)] =$ by the Distributive Law
          $[a \times (b \times c)] + [(a \times (b \times 2)]$

3. (b) $u \times (v + w) =$ by the Distributive Law
          $(u \times v) + (u \times w) =$ by the Commutative law of Multiplication
          $(v \times u) + (u \times w)$

4. No. Consider the value of each side of the given equation when $\ell = 2$, $m = 3$, and $n = 4$.

Section 3.3 (page 80)

1. (a)   not prime   (c)   prime   (e)   not prime   (g)   not prime
    (i)   prime   (k)   not prime   (m)   prime

2. (a) $95 = 5 \times 19$ (e) $100 = 2 \times 2 \times 5 \times 5$
    (g) $1{,}062 = 2 \times 3 \times 3 \times 59$ (k) $1{,}573 = 11 \times 11 \times 13$

3. (a) $1, 5, 19, 95$ (e) $1, 2, 4, 5, 10, 20, 25, 50, 100$
    (g) $1, 2, 3, 6, 9, 18, 59, 118, 177, 354, 531, 1{,}062$
    (k) $1, 11, 13, 121, 143, 1{,}573$

4. Or using the Associative Law of Multiplication, $m \times (k \times \ell) = n$. But $k \times \ell$ is a natural number, so that m must also be a divisor of n. This contradicts the original assumption about m.

Section 3.4 (page 83)

1. $18 = 5 + 13 = 7 + 11$ $22 = 3 + 19 = 5 + 17 = 11 + 11$
    $26 = 3 + 23 = 7 + 19 = 13 + 13$

2. (b)   137 and 139
4. $498 = 2 \times 3 \times 83$ Divisors: 1, 2, 3, 6, 83, 166, 249, 498
    $1 + 2 + 3 + 6 + 83 + 166 + 249 = 510 \neq 498$

Section 3.5 (page 90)

1. 1.   Denote the original number by n
    2.   Add 2 $n + 2$
    3.   Multiply the two numbers $n^2 + 2n$
    4.   Add 7 $n^2 + 2n + 7$

         From the result, subtract 6 $n^2 + 2n + 1$
                                          or $(n + 1)^2$
         Take the square root $n + 1$
         Subtract 1 n

Section 3.6 (page 96)

1. (a) 101111_2 (c) 1234_5 (e) 11011110_2 (g) 2542_8
2. (a) 5_{10} (c) 26_{10} (e) 27_{10} (g) 501_{10}
3. (a) 110100_2 (c) 23_4
4. (a) $12\,\underline{|5,036}$
          $12\,\underline{|419} + 8$
          $12\,\underline{|34} + B$
          $12\,\underline{|2} + A$
          $0 + 2$

Section 3.7 (page 100)

1. $1 = 1_2$ $4 = 100_2$ $7 = 111_2$ $10 = 1010_2$
    $2 = 10_2$ $5 = 101_2$ $8 = 1000_2$
    $3 = 11_2$ $6 = 110_2$ $9 = 1001_2$
2. (h)   26   (i)   42   (j)   63
3. (a) 111_2 (c) 10000_2 (e) 10001_2 (g) 11100_2
4. (a) 1000_2 (c) 101_2 (e) 111_2 (g) 1_2

Section 3.7 (page 102)

1. (a) 110010_2 (c) 10101_2 (e) 110111_2 (g) 1110011_2
2. (a) 1010_2 (c) 111_2 (e) 111_2 (g) 1011_2

ANSWERS FOR CHAPTER 4

Section 4.1 (page 113)

1. (a) $y = 0.10x$ 3. (a) $y = 0.11x$
    (b)   50   (b)   1,100
5. (a) $y = \frac{2}{25}x$ 7. (a) $y = 3 + \frac{1}{2}x$
    (b)   3,125   (b)   24 sq. ft.
9. (a) $y = 25,000 + 1,000x$ 11. $y = 2,000 + 1.10x$
    (b)   26

13. (a) $w = 100 - x$
     (b) $z = 140 + 5x$
     (c) $y = (100 - x)(140 + 5x)$
     (d) \$225

15. (a) $w = 30 - x$
     (b) $y = x(30 - x)$
     (c) 200 sq. ft.

Section 4.2 (page 117)

1. faulty operations   3. incorrect model   5. faulty operations

ANSWERS FOR CHAPTER 5

Section 5.1 (page 130)

1. See Figure B.15.

3. The line parallel to the x-axis passing through the point $(0,2)$.

5. The x-axis.

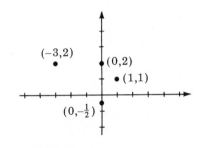

FIGURE B.15

Section 5.2 (page 133)

1. (a) In the graph   (b) In the graph   (c) Not in the graph
    (d) Not in the graph (e) Not in the graph (f) In the graph

3. (a) $(-2,4), (-1,1), (0,0), (1,1), (2,4)$
    (c) $(-2,-3), (-1,0), (0,1), (1,0), (2,-3)$
    (e) $(-2,-5), (-1,-4), (0,-3), (1,-2), (2,-1)$
    (g) $(-2,-8), (-1,-1), (0,0), (1,1), (2,8)$

4. (a) See Figure B.16   (c) See Figure B.17.

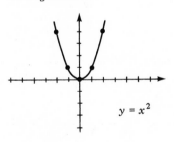

$y = x^2$

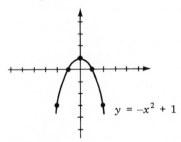

$y = -x^2 + 1$

FIGURE B.16   FIGURE B.17

(e)   See Figure B.18.

(g)   See Figure B.19.

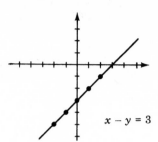

$x - y = 3$

FIGURE B.18

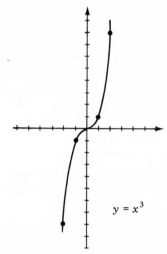

$y = x^3$

FIGURE B.19

Section 5.3 (page 137)

1.   **(a)**   See Figure B.20.

(c)   See Figure B.21.

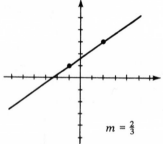

$m = \frac{2}{3}$

FIGURE B.20

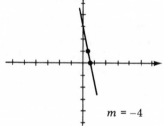

$m = -4$

FIGURE B.21

(e)   See Figure B.22.

(g)   See Figure B.23.

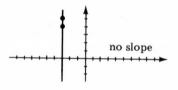

no slope

FIGURE B.22

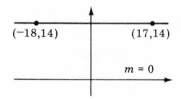

$(-18,14)$ $(17,14)$

$m = 0$

FIGURE B.23

2. (a) See Figure B.24.

(c) See Figure B.25.

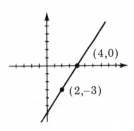

FIGURE B.24

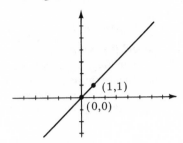

FIGURE B.25

(e) See Figure B.26.

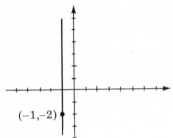

FIGURE B.26

Section 5.3 (page 141)

2. (a) $y - 3 = 2(x - 2)$
    (e) $x = -3$

3. (a) $y = 3x - 2$

(c) $y = -4x + 2$
(g) $y - 14 = 0(x - 7)$ or $y = 14$

(c) $y = -4x + 1$

Section 5.4 (page 143)

1. (a) See Figure B.27.

(c) See Figure B.28.

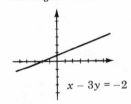

FIGURE B.27

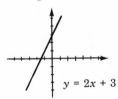

FIGURE B.28

(e) See Figure B.29.

(g) See Figure B.30.

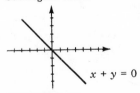

FIGURE B.29

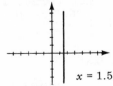

FIGURE B.30

2. **(a)** $\frac{1}{3}$ **(c)** 2   **(e)** −1   **(g)** No slope

Section 5.5 (page 145)

1. **(a) and (b)** See Figure B.31.
   (c) Increase appears to be slow.
   (d) Only those points that have x-coordinates that are integers greater than or equal to zero.

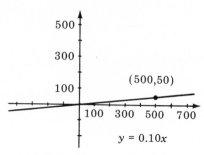

FIGURE B.31

3. **(a) and (b)** See Figure B.32.

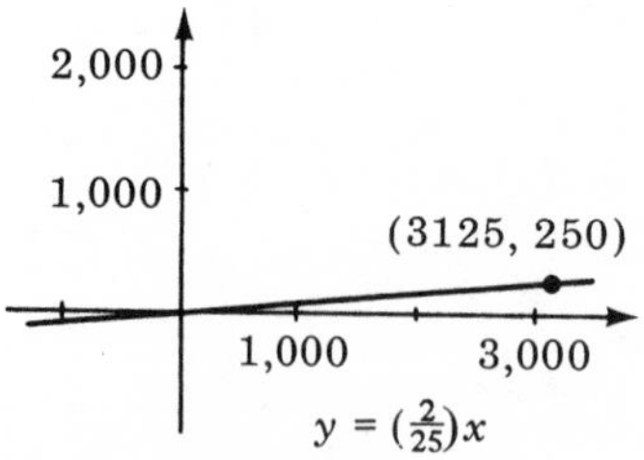

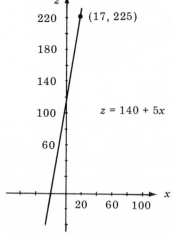

FIGURE B.32

   (c) Same as Exercise 1(d).

7. **(a) and (b)** See Figure B.34.

z = 140 + 5x

FIGURE B.34

   (c) Points that have x-coordinate that are integers, $0 \leqslant x \leqslant 100$.

5. **(a) and (b)** See Figure B.33.

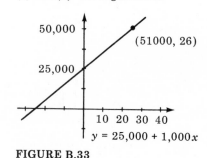

FIGURE B.33

9. **(a) and (b)** See Figure B.35.

r = 150 + 25x

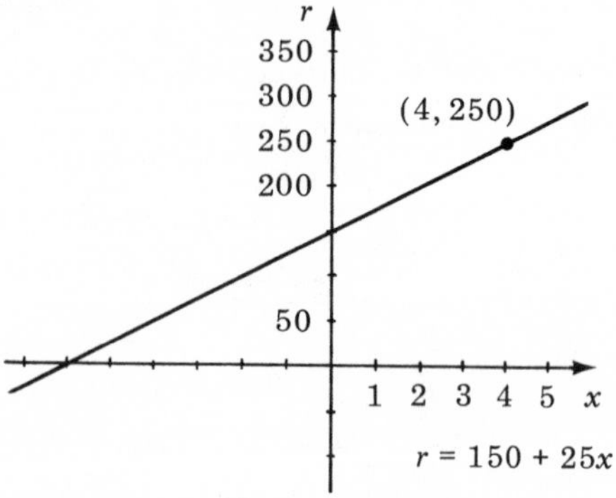

FIGURE B.35

   (c) Points that have x-coordinate that are integers, $0 \leqslant x \leqslant 20$.

Section 5.6 (page 152)

1. (a) $x = 2, y = -1$ (c)   no solution
   (e) $y = x + 4$ or $x = y - 4$

2. (a) $1.5x + 2y = 4{,}775$ (b) $3x + 2y = 5{,}500$
   (c) 483 units of item A $(x = 725/1.5)$
        2,025 units of item B $(y = 2{,}025)$

4. (a) $x + y = 60$ (b) $x = 2y$
   (c) 40 hours on classes; 20 hours on job   (d)   28

6. (a) $x = 1, y = 2, z = 4$

Section 5.7 (page 157)

1. (a)   10   (c)   15   (e) $2\sqrt{2}$

2. (a)   See Figure B.36.   (c)   See Figure B.37.

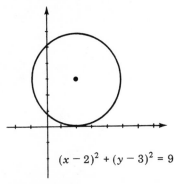

$(x - 2)^2 + (y - 3)^2 = 9$

FIGURE B.36

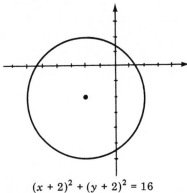

$(x + 2)^2 + (y + 2)^2 = 16$

FIGURE B.37

Section 5.8 (page 164)

1. (a)   See Figure B.38.   (c)   See Figure B.39.

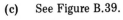

$(2,5)$
$(2,3)$
$y = 1$
$8(y - 3) = (x - 2)^2$

FIGURE B.38

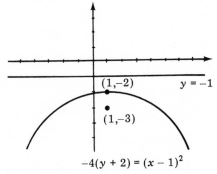

$(1,-2)$ $y = -1$
$(1,-3)$
$-4(y + 2) = (x - 1)^2$

FIGURE B.39

(e)   See Figure B.40.

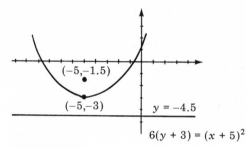

(-5,-1.5)

(-5,-3) $y = -4.5$

$6(y + 3) = (x + 5)^2$

FIGURE B.40.

2.   (a) $-12(y + 1) = (x - 1)^2$ (c) $8(y - 2) = (x - 1)^2$
      (e) $-8y = (x + 3)^2$ (g) $y = x^2$

Section 5.9 (page 168)

1.   (a) $(\frac{1}{4})(y - 1) = (x + 3)^2$. See Figure B.41.

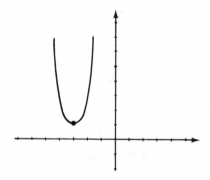

FIGURE B.41

(c) $-8(y + 3) = (x - 16)^2$. See Figure B.42.

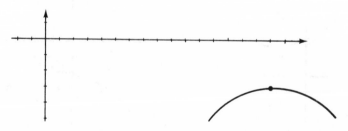

FIGURE B.42

(e) $\frac{1}{4}y = (x + 1)^2$. See Figure B.43.

(g) $-(y - 1.5) = (x + 2.5)^2$. See Figure B.44.

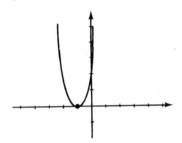

FIGURE B.43

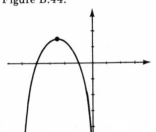

FIGURE B.44

2. (a) $3(y - 2) = (x + 1)^2$; vertex at $(-1,2)$
   (c) $-(y + 4) = (x - 5)^2$; vertex at $(5,-4)$
   (e) $(\frac{1}{8})y = x^2$; vertex at $(0,0)$

3. (c) $y = x^2 - 40x + 435$

Section 5.10 (page 171)

1. (a) 400   **(b)** $325   **(c)** about 81 cents
   (d) See Figure B.45.
   (e) Those points with an integral x-coordinate greater than or equal to zero.

3. (a) $902.50
   (b) 5
   (c) $9.50

5. 15 feet by 15 feet

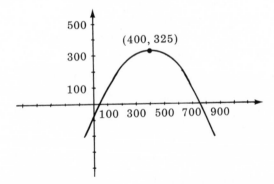

FIGURE B.45

ANSWERS FOR CHAPTER 6

Section 6.1 (page 180)

1. (a) Geometric progression; $r = 3$
   (c) Geometric progression; $r = 5$
   (e) Not a geometric progression

2. (a) $a_3 = 100$ **(c)** $a_7 = 2$ **(e)** $a_5 = -\frac{2}{81}$

3. (a) $3, 12, 48, 192, 768$
   (c) $1, 0.01, 0.0001, 0.000001, 0.00000001$
   (e) $\frac{1}{27}, \frac{1}{9}, \frac{1}{3}, 1, 3$
   (g) $r = 3$: $0.5, 1.5, 4.5, 13.5, 40.5$
         $r = -3$: $-0.5, 1.5, -4.5, 13.5, -40.5$

5. $a_5 = 9,000 \times \frac{16}{81}$ or about 1778

7. $\frac{1}{2} + \frac{1}{4} + \frac{1}{8} = \frac{7}{8}$ quart

Section 6.1 (page 183)

1. (a)   31   (c) $\frac{80}{9} = 8\frac{8}{9}$ (e) $-\frac{13}{4} = -3\frac{1}{4}$

3. See Example 4.   5. $n \times a_1$

Section 6.2 (page 187)

1. $600   3. $203.67   5. $4.61   7. $84.03

9. 1st payment: $150 Interest, $28 Principle
    2nd payment: $149.79 Interest, $28.21 Principle

Section 6.2 (page 190)

3. (a) $1,080   (b)   11%

5. $450   7.   20%   9. $401.79

Section 6.3 (page 194)

1. $2,524.96   2.   (a) $3,166.93   (c) $1,690.25

3. (a)   Amount: $634.87   Interest: $134.87

4. $285.90   6. $197.01

Section 6.3 (page 198)

1. (a) $3,773.34   (c) $1,387.44

2. No; she will have only $9,249.60.

4. There will be $250,040.41 in the account.

Section 6.4 (page 202)

1. 10%   3.   25%   5.   19%

ANSWERS FOR CHAPTER 7

Section 7.1 (page 213)

1. $\frac{31}{60}$ 3.   0.60   5. $\frac{4}{52} = \frac{1}{13}$ 7. $\frac{750}{1000} = \frac{3}{4}$

Section 7.2 (page 218)

2. $4 \times 6 = 24$ 4. $7 \times 5 = 35$

6. (a) $8 \times 7 \times 6 \times 5 \times 4 \times 3 \times 2 \times 1 \times 1 = 40,320$
    (b) $9 \times 8 \times 7 \times 6 \times 5 \times 4 \times 3 \times 2 \times 1 = 362,880$

8. $26 \times 26 \times 10 \times 9 \times 8 \times 7 = 3,407,040$

10. (a) $6 \times 6 \times 6 \times 6 = 1,296$ (b) $1 \times 6 \times 6 \times 6 = 216$
    (c) $216/1,296 = \frac{1}{6}$ (d) $1 \times 1 \times 6 \times 6 = 36$
    (e) $36/1,296 = \frac{1}{36}$

Section 7.2 (page 224)

1. $C(12,3) = 220$ $C(9,1) = 9$ 3. $C(7,3) = 35$

5. $10 \times 9 \times 8 \times 7 = 5,040$

7. $7 \times 7 \times 6 \times 6 \times 5 \times 5 \times 4 \times 4 \times 3 \times 3 \times 2 \times 2 \times 1 \times 1 = 25,401,600$

9.   (a) $C(7,2) \times C(5,3) = 21 \times 10 = 210$
     (b) $C(7,2) \times C(2,2) \times C(3,1) = 63$ elements in the event Bob and Ted are hired; probability is then $\frac{63}{210} = \frac{3}{10}$.

11. $\dfrac{C(4,3) \times C(48,2)}{2,598,960} = \dfrac{4,512}{2,598,960} = 0.0017$

13. $\dfrac{13 \times C(4,3) \times 48 \times 44}{2,598,960} = \dfrac{109,824}{2,598,960} = 0.042$

Section 7.3 (page 229)

1.   (a) $\frac{4}{52} \times \frac{4}{52} = \frac{1}{169}$ 2. (b) $\frac{4}{52} \times \frac{3}{51} \times \frac{2}{50} = \frac{1}{5,525}$

4.   (a) $\frac{7}{12} \times \frac{7}{12} = \frac{49}{144}$ (b) $\frac{7}{12} \times \frac{5}{12} = \frac{35}{144}$

     (c) $\frac{5}{12} \times \frac{7}{12} = \frac{35}{144}$ (d) $\frac{35}{144} + \frac{35}{144} = \frac{35}{72}$

6.   (a) Independent   (b) Dependent

8.   (a) F: the student chosen is from A & S
          E: the student chosen is a boy
          $P(E \cap F) = P(F)P(E|F) = \frac{2}{5} \times \frac{1}{2} = \frac{1}{5}$

     (b) $\frac{10}{50} = \frac{1}{5}$

10. E: a 5 is obtained on first die   12. (a) $\frac{8}{36} = \frac{2}{9}$
     F: a 5 is obtained on second die   (c) $\frac{5}{36} \times \frac{5}{36} = \frac{25}{1,296}$
     Want $P(E \cap F) = \frac{1}{6} \times \frac{1}{6} = \frac{1}{36}$ (e) $\frac{5}{36} \times \frac{25}{36} = \frac{125}{1,296}$

Section 7.4 (page 235)

1. $C(5,3)(\frac{1}{6})^3(\frac{5}{6})^2$ 3. $C(8,4)(\frac{1}{3})^4(\frac{2}{3})^4$ 5. $C(6,2)(\frac{1}{5})^2(\frac{4}{5})^4$

7.   (a) $C(5,3)(\frac{1}{2})^5$ (b) $C(5,3)(\frac{1}{2})^5 + C(5,4)(\frac{1}{2})^5 + C(5,5)(\frac{1}{2})^5$

9. $1 - [C(10,0)(0.10)^0(0.90)^{10} + C(10,1)(0.10)^1(0.90)^9]$

Section 7.5 (page 237)

1.   cup B; $\frac{1}{2} \times \frac{1}{2} = \frac{1}{4}$ cup C, cup D: $\frac{1}{2} \times \frac{1}{2} \times \frac{1}{2} = \frac{1}{8}$

3. $1 - \frac{1}{2} = \frac{1}{2}$

5. $S = \{A, B, C, D\}$
     $P(A) = \frac{1}{2}, P(B) = \frac{1}{4}, P(C) = P(D) = \frac{1}{8}$

7.   (b) $C(3,2)(\frac{3}{8})^2(\frac{5}{8})^1$

ANSWERS FOR CHAPTER 8

Section 8.1 (page 244)

1.   See Figure B.46.

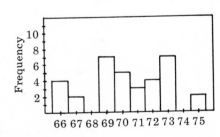

FIGURE B.46

Section 8.2 (page 247)

1. (a)   Mean: 27   Median: 26
3. (a)   Type I: 42.35   Type II: 41.74   (b)   Type I
   (c)   Type I: 42.4   Type II: 42.1   (d)   Type I
5. Mean: 9.5   Median: 9.6

Section 8.3 (page 255)

1. Mean: 72; Standard Deviation: 7; Interquartile Range: $80 - 64 = 16$; Range: 20
3. (a)   Within one standard deviation on the first, not on the second.
   (b)   Better on the first.
5. Mean: 70.4; Standard Deviation: $\sqrt{216.04/34}$ or approximately 2.5; Interquartile Range: $73 - 69 = 4$; Range: 9

Section 8.4 (page 263)

1. Type A indicated superior.
3. Neither fertilizer superior.
5. Domestic brand indicated superior.

Section 8.5 (page 273)

1. (a)   0.385   (c) $0.495 - 0.433 = 0.062$
   (e) $0.341 + 0.477 = 0.818$ (g) $0.433 + 0.5 = 0.933$
   (i) $0.023 + 0.023 = 0.046$
2. (a)   1.96   (b) -1.64

Section 8.5 (page 276)

1. (a) $0.499 - 0.341 = 0.158$ (b)   0.226
   (c) $0.419 + 0.477 = 0.896$ (d) $0.5 - 0.499 = 0.001$
   (e) $0.5 + 0.155 = 0.655$
3. $0.192 + 0.192 = 0.384$
5. (a) $0.5 - 0.477 = 0.023$ (b)   0.046
6. (a)   73.72   (c)   69.1

Section 8.5 (page 280)

1. Claim rejected ($z = 2$)   3. Claim accepted ($z = -1.5$)

ANSWERS FOR CHAPTER 9

Section 9.1 (page 290)

1. "$\times$," "1," and "$1/a$" respectively.
2. (b)   Multiplication: Axiom 4 not satisfied.
          Addition: Axiom 1 not satisfied
                     Axiom 3 not satisfied
                     Axiom 4 not satisfied
3. Axiom 1: For any rational number a, a divided by 0 is not another natural number—in fact, it has no meaning.

Axiom 2: For example, $2 \div (3 \div 4) \neq (2 \div 3) \div 4$
Axiom 3: 1 cannot be the identity because of 0. 1 divided by 0 has no meaning.
Axiom 4: If there is no identity, there can be no inverses.

Section 9.2 (page 298)

1. (a) $c * d = b$ (b) $b * (a * b) = c$ (d) $a^{-1} = a$

2. (a)   R3 * D2 = D2   (b)   R1 * (D2 * R2) = D2   (d) $(R1)^{-1} = R2$

4.   0 has no inverse.

6. (a)   (i) $\begin{bmatrix} 1 & 7 \\ 0 & -10 \end{bmatrix}$ (iii) $\begin{bmatrix} 0 & 0 \\ 0 & 0 \end{bmatrix}$

    (c)   (i) $\begin{bmatrix} -2 & 2 \\ 5 & -7 \end{bmatrix}$ (iii) $\begin{bmatrix} 1 & 2 \\ -2 & -1 \end{bmatrix}$

8. (a)   See Table B.1.

TABLE B.1

	1	-1	i	$-i$
1	1	-1	i	$-i$
-1	-1	1	$-i$	i
i	i	$-i$	-1	1
$-i$	$-i$	i	1	-1

10.   In ordinary arithmetic, $1 + 1 = 2$; in binary arithmetic, $1 + 1 = 10$; in modulo 2 arithmetic, $1 + 1 = 0$. The first meaning is presumed when no further explanation is given.

Section 9.3 (page 305)

1. 1.   This statement is the assumption of the conditional part of the theorem.
    2.   Because of the equality expressed in Statement 1, d^{-1} operating on $d * f$ is equal to d^{-1} operating on $d * g$.
    3.   Because of Axiom 2, the terms on both sides of Statement 2 can be regrouped.
    4.   By Axiom 4, $d^{-1} * d$ in both sides of Statement 3 is the identity element, e.
    5.   By Axiom 3, $e * f$ in Statement 4 is the same as f, and $e * g$ is the same as g.
    6.   This is the statement of the theorem. By assuming the conditional part in Statement 1, we were able to arrive at the conclusion of the theorem in Statement 5 by means of the axioms.

2.

Proof	*Analysis*
1. Assume that $g = f * d^{-1}$.	This statement assumes the conditional part of the statement of the theorem.
2. Then, $g * d = (f * d^{-1}) * d$.	Because of the equality in Statement 1, g operating on d is the same as $f * d^{-1}$ operating on d.
3. Therefore, $g * d = f * (d^{-1} * d)$.	The terms on the right side of Statement 2 can be regrouped by Axiom 2.
4. Also, $g * d = f * e$.	By Axiom 4, $d^{-1} * d$ in Statement 3 is the same as the identity e.

5. Hence, $g * d = f$.

    By Axiom 3, $f * e$ in Statement 4 is the same as f.

6. Consequently, if $g = f * d^{-1}$, then $g * d = f$.

    This is the statement of the theorem. From the assumption of the conditional part in Statement 1, we obtained the conclusion in Statement 5 by means of the axioms.

4. 3. $f * (d * d^{-1}) = g * (d * d^{-1})$
    4. $f * e = g * e$
    5. $f = g$
    6. If $f * d = g * d$, then $f = g$.

ANSWERS FOR CHAPTER 10

Section 10.1 (page 319)

1. (a) See Figure B.47.

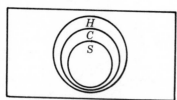

C: Congressmen
H: Honest people
S: Senators
Valid

FIGURE B.47

(c) See Figure B.49.

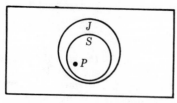

S: Students who work hard
J. People who get good jobs
P: Paul
Valid

FIGURE B.49

(b) See Figure B.48.

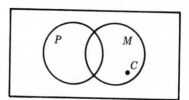

M: Magazines
P: Publications worth reading
C: "Collegean"
Invalid

FIGURE B.48

(f) See Figure B.50.

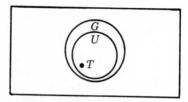

U: Understanding parents
G: Good parents
T: Tom's mother
Valid

FIGURE B.50

Section 10.2 (page 322)

1. (a) $\sim q$
    (d) $q \wedge \sim r$

    (b) $p \wedge r$
    (e) $(p \vee q) \vee r$

    (c) $(\sim p) \to (\sim r)$

3. (a) If exercising is essential for good health and eating too much causes strain on the heart, then good health retards old age.

(b) Either exercising is not essential for good health or good health does not retard old age.

(c) If exercising is essential for good health, then good health retards old age, and if good health retards old age, then eating too much causes strain on the heart.

(d) Eating too much does not cause strain on the heart or good health retards old age.

Section 10.3 (page 325)

1.

p	q	$\sim q$	$p \wedge (\sim q)$
T	T	F	F
T	F	T	T
F	T	F	F
F	F	T	F

2.

p	q	$p \vee q$	$\sim(p \vee q)$
T	T	T	F
T	F	T	F
F	T	T	F
F	F	F	T

3.

p	q	$\sim p$	$\sim q$	$(\sim p) \wedge (\sim q)$
T	T	F	F	F
T	F	F	T	F
F	T	T	F	F
F	F	T	T	T

6.

p	q	$p \to q$	$\sim(p \to q)$
T	T	T	F
T	F	F	T
F	T	T	F
F	F	T	F

9.

p	q	r	$p \to q$	$r \wedge (p \to q)$
T	T	T	T	T
T	T	F	T	F
T	F	T	F	F
T	F	F	F	F
F	T	T	T	T
F	T	F	T	F
F	F	T	T	T
F	F	F	T	F

Section 10.3 (page 326)

1.

p	q	$p \to q$	$\sim p$	$(\sim p) \vee q$
T	T	T	F	T
T	F	F	F	F
F	T	T	T	T
F	F	T	T	T

Columns 3 and 5 verify equivalence.

3.

q	$\sim q$	$\sim(\sim q)$
T	F	T
F	T	F

Columns 1 and 3 verify equivalence.

5. **(a)** 5 **(c)** 6 **(e)** 3

Section 10.4 (page 331)

1. **(a)** p and q closed, r closed, or p, q, and r closed.
(c) r, p, and q closed, or r and $\sim q$ closed.
(e) never

2. **(a)** $(p \wedge q) \vee r$ **(c)** $r \wedge [(p \wedge q) \vee (\sim q)]$ **(e)** $p \wedge (\sim p)$

3. **(a)**   See Figure B.51.

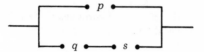

FIGURE B.51

(c)   See Figure B.52.

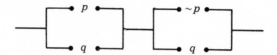

FIGURE B.52

(e)   See Figure B.53.

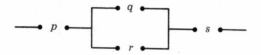

FIGURE B.53

Section 10.4 (page 335)

1.

p	q	r	$q \wedge r$	$p \wedge (q \wedge r)$	$p \wedge q$	$(p \wedge q) \wedge r$
T	T	T	T	T	T	T
T	T	F	F	F	T	F
T	F	T	F	F	F	F
T	F	F	F	F	F	F
F	T	T	T	F	F	F
F	T	F	F	F	F	F
F	F	T	F	F	F	F
F	F	F	F	F	F	F

Columns 5 and 7 verify equivalence.

3. **(a)** $p \wedge (q \vee \sim q)$
    p by 13

    (c) $(p \vee q) \wedge p$
    $p \wedge (p \vee q)$ by 7
    p by 15

    (e) $p \wedge [q \wedge (q \vee \sim p)]$
    $p \wedge q$ by 15

4. **(a)** $(p \wedge q) \vee (\sim p \wedge q)$
    $(q \wedge p) \vee (q \wedge \sim p)$ by 7
    $q \wedge [p \vee \sim p]$ by 11
    q by 13

    (c) $p \wedge [q \wedge (\sim q \vee r)]$
    $p \wedge [(q \wedge \sim q) \vee (q \wedge r)]$ by 11
    $p \wedge [(q \wedge r) \vee (q \wedge \sim q)]$ by 8
    $p \wedge (q \wedge r)$ by 14

Section 10.5 (page 338)

1. p: Paul studies programming $p \to q$
   q: Paul could use the computer. $\dfrac{\sim q}{}$
                                          $\therefore \quad \sim p$

p	q	$p \to q$	$\sim q$	$\sim p$	$(p \to q) \wedge (\sim q)$	$[(p \to q) \wedge (\sim q)] \to (\sim p)$
T	T	T	F	F	F	T
T	F	F	T	F	F	T
F	T	T	F	T	F	T
F	F	T	T	T	T	T

Argument valid.

4. p: Elected representatives serve their constituents.
   q: Elected representatives lose their jobs.

   $p \vee q$
   $\dfrac{\sim q}{}$
   $\therefore \quad \sim p$

p	q	$p \vee q$	$\sim q$	$\sim p$	$(p \vee q) \wedge (\sim q)$	$[(p \vee q) \wedge (\sim q)] \to (\sim p)$
T	T	T	F	F	F	T
T	F	T	T	F	T	F
F	T	T	F	T	F	T
F	F	F	T	T	F	T

Argument invalid.

5. p: Joan cares for people. $(\sim p) \vee q$
   q: Joan studies sociology. $\dfrac{p}{}$
                                    $\therefore \quad q$

p	q	$\sim p$	$(\sim p) \vee q$	$(\sim p \vee q) \wedge p$	$[(\sim p \vee q) \wedge p] \to q$
T	T	F	T	T	T
T	F	F	F	F	T
F	T	T	T	F	T
F	F	T	T	F	T

Argument valid.

6.

p	q	$p \to q$	$\sim q$	$(p \to q) \wedge (\sim q)$	$\sim p$	$[(p \to q) \wedge (\sim q)] \to (\sim p)$
T	T	T	F	F	F	T
T	F	F	T	F	F	T
F	T	T	F	F	T	T
F	F	T	T	T	T	T

Argument valid.

8.

p	q	r	$p \to q$	$q \to r$	$p \to r$	$(p \to q) \land (q \to r)$	$[(p \to q) \land (q \to r)] \to (p \to r)$
T	T	T	T	T	T	T	T
T	T	F	T	F	F	F	T
T	F	T	F	T	T	F	T
T	F	F	F	T	F	F	T
F	T	T	T	T	T	T	T
F	T	F	T	F	T	F	T
F	F	T	T	T	T	T	T
F	F	F	T	T	T	T	T

Argument valid.

10.

p	q	$p \to q$	$(p \to q) \land p$	$[(p \to q) \land p] \to q$
T	T	T	T	T
T	F	F	F	T
F	T	T	F	T
F	F	T	F	T

Argument valid.

Section 10.6 (page 340)

1.

p	q	$p \lor q$	$\sim p$	$\sim q$	$(p \lor q) \land (\sim p)$	$[(p \lor q) \land (\sim p)] \to q$	$[(p \lor q) \land (\sim q)] \to p$
T	T	T	F	F	F	T	T
T	F	T	F	T	F	T	T
F	T	T	T	F	T	T	T
F	F	F	T	T	F	T	T

Columns 7 and 8 verify Rule of Inference.

3.

p	q	$p \land q$	$(p \land q) \to (p \land q)$	$q \land p$	$(q \land p) \to (p \land q)$
T	T	T	T	T	T
T	F	F	T	F	T
F	T	F	T	F	T
F	F	F	T	F	T

Columns 4 and 6 verify Rule of Inference.

4.   (a)   4   (c)   7   (e)   3   (g)   2

Section 10.7 (page 344)

1.   1   P2, P1, R3   3.   1   P2, R6
     2   S1, ES   2   S1, P1, R1

5.   1   P1, R5   or   1   P1, R5
     2   S1, P2, R1   2   P2, P3, R2
     3   S2, P3, R1   3   S1, S2, R1

7.   1   P3, ES
     2   S1, P1, R4
     3   S2, P2, R1

9. 1 P3, R5
    2 S1, P1, R1
    3 S1, ES
    4 S3, P2, R3
    5 S2, S4, R7
    6 S5, ES

11. 1 P4, P3, R1
     2 S1, P1, R4
     3 S2, P2, R2

Section 10.7 (page 347)

1. (a) 1 $p \lor q$ P1, R6
         2 r S1, P2, R1
    (c) 1 $(\sim p) \land (\sim q)$ P1, ES
         2 $\sim p$ S1, R5
         3 $\sim s$ S2, P2, R3
    (e) 1 $p \land (\sim q)$ P1, ES
         2 p S1, R5
         3 $p \lor q$ S2, R6
         4 s S3, P2, R1

2. (b) p: We are concerned for ourselves.
         q: We are concerned for future generations.
         r: We pay heed to ecology.

	Statement Number	Valid Statement	Reasons
$(p \lor q) \to r$	1	$\sim(p \lor q)$	P1, P2, R3
$\sim r$	2	$(\sim p) \land (\sim q)$	S1, ES
∴ $\sim q$	3	$\sim q$	S2, R5

Section 10.8 (page 350)

1. 1 Assumption
    2 P1, P2, R3
    3 S2, ES
    4 S1, ES
    5 S3, S4, R4
    6 S5, PI

3. 1 Assumption
    2 S1, R6
    3 S2, P1, R1
    4 S3, P2, R1
    5 S4, PI

5. 1 Assumption
    2 S1, P1, R1
    3 S2, R5
    4 S3, ES
    5 S4, P2, R4
    6 S5, PI

Section 10.8 (page 353)

1. 1 Assumption
    2 S1, P1, R1
    3 S2, P2, R7
    4 S3, PC

3. 1 Assumption
    2 S1, P3, R3
    3 S2, P2, R3
    4 P1, R5
    5 S3, S4, R7
    6 S5, PC

5. 1   Assumption
    2   S1, P1, R1
    3   P2, ES
    4   S3, R5
    5   S2, S4, R7
    6   S5, PC

Section 10.9 (page 359)

2. 1   Assumption
    2   S1, ES
    3   S2, A3, R1
    4   S3
    5   S4, A1, R7
    6   S5, PC

4. Axiom 1

6. Eliminate Axiom 1 and add the following axiom:

> For every three members of B, a, b, and c, at least one of (abc), (bac) or (acb) must hold.

ANSWERS FOR APPENDIX A

Appendix A (page 367)

(Because the conversion factors are approximations, your solutions may vary somewhat, depending on your method of solution.)

1. 75.8 ℓ	3. 340.2 g	5. 45.5 m	9. 99.2 mi
11. 148.2 acres	13. 0.53	15. 4 ℓ	

17. Not quite a cup. (0.90 cup)

Appendix A.1 (page 370)

1. About 235 yd^3

3. 5 qt, if coverage is not to exceed specifications.

5. $15.00

index

Pages in bold print refer to beginning of chapters or sections.

Photo Acknowledgements

Page

xiv	The Royal Irish Academy, Dublin.
2	The Granger Collection.
14	The Board of Trinity College, Dublin.
36	The Toledo Museum of Art, Toledo, Ohio. Gift of the Macomber Family in memory of C. Reynolds Macomber, 1971.
37	Brooks Caldwell, The Stock Market.
39	Wolff-Leavenworth Collection, George Arents Research Library at Syracuse University (upper right).
39	Culver Pictures, Inc. (bottom left)
39	The Bettmann Archive, Inc. (bottom right)
66	Photo courtesy of IBM.
69	Reprinted by permission of Hubert C. Kennedy.
104	Escher Foundation, Collection Haags Gemeentemuseum, The Hague
121	The Bettmann Archive, Inc.
124	Photo courtesy of Missouri Botanical Garden.
126	Brown Brothers
173	Photo by Arteaga Photos—St. Louis, Mo.
174	Photograph by Charles Eames. Courtesy of IBM.
175	Photo by Bill Hartough.
208	Bittinger/Crown, *Finite Mathematics: A Modeling Approach*, © 1977, Addison-Wesley, Reading, Massachusetts, pp. 229. Reprinted with permission.
240	Courtesy Museum of Science and Industry, Chicago.
286	Photo by Andrew C. Kasiske.
287	Brown Brothers.
311	Photo by Bill Hartough.
314	Drawing by John Tenniel.